Fundamental Theories of Physics

Volume 207

The international monograph series "Fundamental Theories of Physics" aims to stretch the boundaries of mainstream physics by clarifying and developing the theoretical and conceptual framework of physics and by applying it to a wide range of interdisciplinary scientific fields. Original contributions in well-established fields such as Quantum Physics, Relativity Theory, Cosmology, Quantum Field Theory, Statistical Mechanics and Nonlinear Dynamics are welcome. The series also provides a forum for non-conventional approaches to these fields. Publications should present new and promising ideas, with prospects for their further development, and carefully show how they connect to conventional views of the topic. Although the aim of this series is to go beyond established mainstream physics, a high profile and open-minded Editorial Board will evaluate all contributions carefully to ensure a high scientific standard.

More information about this series at https://link.springer.com/bookseries/6001

Erhard Scheibe

The Reduction of Physical Theories

A Contribution to the Unity of Physics
Part 1: Foundations and Elementary Theory

Springer

Erhard Scheibe
Deceased
Hamburg, Germany

Translated by
Brigitte Falkenburg
Institut für Philosophie
und Politikwissenschaft
TU Dortmund
Dortmund, Germany

Gregg Jaeger
Natural Sciences and Mathematics
Boston University
Boston, MA, USA

ISSN 0168-1222 ISSN 2365-6425 (electronic)
Fundamental Theories of Physics
ISBN 978-3-662-64998-5 ISBN 978-3-662-65000-4 (eBook)
https://doi.org/10.1007/978-3-662-65000-4

Translation from the German language edition: "Die Reduktion physikalischer Theorien. Ein Beitrag zur Einheit der Physik Teil 1: Grundlagen und elementare Theorie" by Erhard Scheibe, © Springer-Verlag Berlin Heidelberg 1997. Published by Springer. All Rights Reserved.

This Springer imprint is published by the registered company Springer-Verlag GmbH, DE part of Springer Nature.
The registered company address is: Heidelberger Platz 3, 14197 Berlin, Germany

...und in den Wäldern des Amazonenflusses wie auf dem Rücken der hohen Anden erkannte ich, wie von einem Hauche beseelt von Pol zu Pol nur ein Leben ausgegossen ist in Steinen, Planzen und Tieren und in des Menschen schwellender Brust.

Alexander von Humboldt to Caroline von Wolzogen, May 14, 1806

To Carl Friedrich von Weizsäcker
in gratitude and adoration

Translators' Preface

Die Reduktion physikalischer Theorien is the *opus magnum* of the German philosopher of science Erhard Scheibe (1927–2010). For more than two decades, only fragments of this outstanding theory of reduction in physics were accessible in English, through a series of his articles with general considerations and case studies included in the collection *Between Rationalism and Empiricism. Selected Papers in the Philosophy of Physics* (2001). We are glad to present now the translation of the first volume of his comprehensive study *Die Reduktion physikalischer Theorien. Ein Beitrag zur Einheit der Physik. Teil I: Grundlagen und elementare Theorie* (1997). It explains the philosophical background and the logical foundations of Scheibe's theory of reduction. The second volume *Die Reduktion physikalischer Theorien. Ein Beitrag zur Einheit der Physik. Teil II: Inkommensurabilität und Grenzfallreduktion* (1999), the translation of which is also in preparation, deals with incommensurability and the reduction of the classical theories to special relativity, general relativity, and quantum mechanics.

Scheibe's approach focuses on reduction within physics, and it does so on a highly sophisticated formal level. Nevertheless, it is also relevant for a deeper understanding of reduction in general that takes scientific pluralism into account. As Scheibe himself emphasizes in the *Preface* to the present volume, the motivation to develop his approach was the question of how scientific progress may be understood in the face of Kuhn's and Feyerabend's attacks against Logical Empiricism and Critical Rationalism. Scheibe's answer was a rational reconstruction of the architectonics of physics based on a detailed analysis of the inter-theoretical relations between the laws of classical mechanics, special and general relativity, and quantum mechanics. This rational reconstruction aims, in particular, at clarifying the logical relations between classical mechanics and quantum mechanics, which in Scheibe's view is the hardest case of incommensurability. The inter-theoretical relations of physics show that reduction has many faces. There is no unique, all-encompassing concept of reduction. The case studies discussed in the book reveal that there are several stronger and weaker kinds of reductions—different kinds of exact reductions, approximate reductions, and partial reductions, which quite often also need to be successively combined in order to reduce a theory to its successor. The resulting theory of reduction is *synthetic* (rather than analytic) and *recursive*. It is open for adding new kinds of reduction, and it demonstrates that theoretical reduction has its limits already in physics—not to speak of the possibility to reduce higher-level sciences such as biology and chemistry to physics, for which the situation is expected to be much worse. In a certain sense, Scheibe's approach takes the disunity of science into account, which has been emphasized by Nancy Cartwright in particular, at the time when Scheibe developed his theory of reduction in physics. His approach, however, shows that the plurality of theories or theory fragments that coexist in the practice of physics is neither irrational nor arbitrary. It gives systematic access to the web of inter-theoretical relations that make up the architectonics of physics after the quantum revolution, due to which the unity of physics was lost.

Scheibe's theory of reduction differs in several respects from the predominant tendencies of philosophy of science in the second half of the twentieth century. Based on detailed case studies as well as the different conceptions of reduction presented by Nagel and other leading philosophers of science in the 1950s, Scheibe develops a pluralistic concept of reduction, which limits the prospects of a reductionist scientific worldview already from its basis in physics. He is able to reconcile Kuhn's insight into the incommensurability of theories with the rational reconstruction of their inter-theoretical relations. For the rational reconstruction, he chooses a semantic approach, which crucially differs from the contemporary empiricist attempts to reconstruct the architectonics of physics from its empirical basis. Scheibe's reconstruction of physics introduces physical objects in a very broad sense, from space-time points to electrons, as basic elements into set theory, taking up Zermelo's conception of individuals (*Urelemente*). And his formal tool is not naïve set theory, but Bourbaki's structuralism. This work illustrates clearly through its novel and detailed treatment of the reduction problem in physics, which has long served as the proving ground for understanding science in general, how the success or failure of future attempts to reduce theories meant to describe different physical scales or subject areas can be thoroughly assessed by taking into account the specific character of the theories under consideration.

In the translation, we have corrected all obvious errors we found in the German edition, relying on an extremely helpful error list provided by Johannes Mierau. For better reading and more clarity concerning cross-references between the chapters, we changed the equation numbers to consecutive numbering within each chapter. All footnotes are ours. We added them to explain Scheibe's approach and to indicate differences between the announcements made in Vol. I and the final contents of Vol. II. In the footnotes, we also make some handwritten comments available which we found in Scheibe's own copy of the book.

In the Appendix, we added the table of contents of Vol. II, which corresponds to Part II of the German edition published in 1999. It replaces Scheibe's original plan of the contents of Part II, as given in 1997 at the end of the German edition of the present book. In addition, we included Brigitte Falkenburg's memorial article *The Legacy of Erhard Scheibe*, which was first published in the *Journal for General Philosophy of Science* (2011), and the complete list of Erhard Scheibe's works taken from there.

We would like to thank Angela Lahee from Springer for supporting the overdue translation of this important book, Johannes Mierau for his careful proofreading of our translation, and Paul Busch for his engagement with the project in its early stages before his recent, untimely passing.

Berlin, Germany
Boston, USA
January 2022

Brigitte Falkenburg
Gregg Jaeger

Preface

My motive for considering the subject of this book was the ways in which T. Kuhn and P. Feyerabend attacked the philosophy of science of logical positivism and of Popper. Even then, in the early seventies, I myself still maintained a basically reductionist attitude, perhaps due to the early influence of my philosophy teacher C. F. von Weizsäcker. The idea of devoting an entire book to the subject emerged in the mid-eighties. From then on, I looked for opportunities to focus on this work. The first came in the form of a visiting professorship at the University of California at Irvine during the Winter semester of 1987. There, at the invitation of Prof. K. J. Lambert, I held lectures entitled "Explanation, Reduction, Progress". A second opportunity came through a 10-month fellowship at the Institute for Advanced Study of Berlin in 1987–1988. The relatively long period of exemption from teaching duties even gave hope that I would already be able to finish the book in Berlin. This by no means happened. Only two years later, at a colloquium held in Pittsburgh in honor of A. Grünbaum, could I present the main idea on which today's book is based. The research semester of 1990–1991, partly spent in Delhi as a guest of Prof. G. Pandit, further served the formation of its conception. But the detailed work had still to wait until my retirement.

Of those with whom I occasionally conversed, I would like in addition to gratefully mention J. Ehlers and H.-I. Schmidt. From the beginning and repeatedly until his death, my path brought me together with L. Krüger. With G. Ludwig, I had the exchange of ideas of most far-reaching scope and content on relevant issues. The latter made such a lasting impression on me because it took place not only in oral discussions, as is usual today, but also in large part through an extensive correspondence.

A special concession from Springer is that the book can be published in two parts—the first now and a second within a year. My colleague from Heidelberg, Prof. W. Beiglböck, has followed the progress of my work with continued interest and has arranged for the possibility of bringing my thoughts to a broader public.

The text of the book was written by me after my retirement—without a secretary and on a PC. I could not have done it if my sons had not introduced me to such a device and had not also had an impact on the production process itself.

To all named, my heartfelt thanks.

Hamburg, Germany
September 1996

Erhard Scheibe

Introduction

> Denn so laut er auch die
> Unerschütterlichkeit seines
> Systems proklamieren mochte,
> gerade hinter seinen
> bestimmtesten Versicherungen barg sich
> der quälendste Zweifel. Alle Systeme
> sind gefallen, sagte er zu sich selbst,
> und vor jeder neuen Debatte beschlich
> ihn die Vorstellung: wenn nun jetzt
> dein Bau zusammenstürzte!
>
> *Fontane, Vor dem Sturm*

This first part of this book is dedicated to the reduction of physical theories. Instead of 'reduction', one could also say the *explanation* of physical theories (through others). Because, even though one does not want to deny that this distinction can be made, it does not suggest itself in the context of the theory to be developed here. Since reductions of several theories to a single one are carried out as a unification of the former by the latter, the book is also a contribution to the unity of physics. Finally, it will matter that many examples of reductions which have actually occurred in the history of physics are at the same time examples of empirical progress in theory formation. "Reduction, Progress and Unity of Physics" would be another possible title for the book. But the emphasis of the presentation will be on the reductions themselves.

One knows them from science, first of all as the 'grand' reductions of entire disciplines, such as biology to physics, or better, as the *problems* of such reductions. These never actually manage to succeed. The pre-Socratic philosophers already established broadly reductionist theses when they claimed that the first principle of all existence is one or the other material, for example, water (Thales) or something more abstract, such as the 'indefinite' of Anaximander. These approaches do not seem to have progressed very far. Something else was already the case with the atomistic thesis, according to which all matter was built out of immutable atoms, the different positions and movements of which should explain the multiplicity of phenomena (including sensation!). This doctrine of atomism, going back to Leucippus, had already in Antiquity developed a certain tradition (Democritus, Epicurus, and others). Above all, it has been taken up again in modern times (Gassendi, Hobbes, and others). Incorporating the doctrine of primary and secondary qualities, it has led to the first large reduction program of the *mechanization* of the natural sciences.

The reduction to mechanical (and, a fortiori, geometrical) qualities in the sense of modern physics and chemistry could be atomistic, but it needn't be. Newton's mechanics had established such a stable and successful conceptual system that the mere mechanization of processes was already considered of sufficient explanatory power. This was in particular true for the nineteenth century when, especially in England, the explanatory success of mechanical 'models' was cherished. But the futile attempts to understand gravity as a mechanism and the successes of the field concept in electrodynamics, as well as the incompatibility of the latter with Newtonian mechanics, eventually led to abandoning the mechanistic program. Only the atomistic branch remained excluded, which, under Boltzmann's and Maxwell's hands, had partial explanatory success in describing the thermodynamics of gases. This development, however, proceeded through replacing classical mechanics by quantum mechanics. The latter made for the first time an atomic physics possible which complied with the methodological standards achieved in the meantime, making parts of chemistry, in particular, the periodic table of elements and chemical bonding, accessible for atomistic explanations. But to summarize these successes as a reduction of the whole of chemistry to physics would certainly still be a euphemism. Accordingly, the physicists who had achieved these successes preferred to speak of a 'fusion' of physics and chemistry, thereby leaving the intrinsic importance of the latter untouched.

The reduction of biology to chemistry and physics has also become a favored topic of discussion. In the middle of the previous century, scholars such as Helmholtz and E. Du Bois-Reymond explicitly established physiology as a scientific bridge from biology to physics. The Darwinian theory of descent was a physicalization of what had been a predominantly static perspective of biological species, an evolutionary way of thinking that was taking its place. In our century, molecular biology then contributed significantly to anchoring the physico-chemical conception of evolution and of the subject area of biology in general. Again, however, to summarize these results as a reduction of biology to physics and chemistry would serve to obscure the situation rather than to clarify it. As already in the case of the mechanization of physics, such a summary of the hard-won successes of molecular biology would again put at risk their scientific standards reached in between, bringing a reductionist terminology into play which has not yet been sufficiently thought through. It is far too unclear, in such unqualified reductive claims, what the reduction counterparts are here, not to mention what is to be understood by the respective reduction itself. Even the generalized analysis of phenomena—theory formation—is already a kind of reduction of phenomena to a theory. One can reduce languages to other languages, theories to other theories, and objects to other objects of which they are made up—all this flows uncontrolled in the general discussion of these reductions and even leaves what is at issue barely recognized.

One possibility for circumventing these difficulties, or at least diminishing them, consists in completely excluding for the time being the grandiose perspectives which are given by the reduction of chemistry to physics, biology to chemistry, or even psychology to physics. Indeed, it turns out that this does not mean that the issue of reduction itself goes overboard. Rather, the reduction business, provided one does not make it too restrictive, is already generating large revenues *within physics*, and its precise treatment is at least possible if physics is included in the undertaking as a certain *rational reconstruction*. These, in any case, are the two limitations that are made in this book if one still does not specifically add the third, which is that it will predominantly be about *theory reduction*, i.e., reductions that reduce theories to other theories. Together with the other two restrictions, the achievements regarding reduction will hence be substantially based on a reconstruction of the concept of physical theory. Whether, in this form, they throw a light on those great questions of the connections all sciences have with each other, and what light this would be, is a *cura posterior* which remains undiscussed in this book.

In this framework, restricted by content to physics and by method to reconstructionism, what is genuinely new that awaits the reader in this book is, on the one hand, a *theory of reduction* not yet found in the literature (or, to be more precise, a meta-theory of reduction of physical theories) and, on the other hand, an especially detailed *exemplification of the reduction problem*. Theory and exemplification are indeed here to be seen independently, which in the two-part division of the work comes particularly to the fore. In Vol. I, the theory stands in the foreground and is introduced and explained through simple examples from physics. The explanation of the Galilean law of falling bodies by Newtonian gravitational theory is perhaps typical of the examples relied upon here. The exemplification given in Vol. I certainly receives at least quantitative support in the large number of examples provided, compared to what is usual in the philosophy of science. In Vol. I, however, first and foremost, a general theory of reduction is developed. Its novelty is as follows. It replaces the usual attempt to center such a theory around *one general* notion of reduction, which holds for all instances, by the approach of finding distinct kinds of reduction, which are as *specific* as possible and have no proper decomposition, and from which further reduction types are obtained by combing them (i.e., by composition). In this way, the analytic explication of a prior ‘pre-scientific’ concept of reduction is replaced by a synthetic (or, recursive) construction of the reduction concept based on the iteration of reductions. For this reduction concept, however, it must remain unclear for the time being what distinguishes its elementary, i.e., no longer genuinely decomposable reductions.

This is, in fact, already the whole idea on which this new theory is built. Its novelty comes primarily to the fore by its taking on a character different from that which a theory based on the usual explication method would take on (or has done, to the extent that it exists at all—see the literature mentioned in the introduction to Chap. 1). In particular, the analysis of the examples turns out to be much more productive because each example that is at all recognized as such immediately appears as one or the other elementary reduction, or as a combination of the two, whereas following the usual method, it would be identified as merely an instance of the general notion of reduction, just like any other example. In all remaining respects, the synthetic concept of reduction preferred here is constructed without prejudice to the general *guiding notion* that, in principle, a physical theory is made dispensable or needless or redundant by the very theory to which it is reduced.

Vol. II reverses the relationship of theory and example: The examples, now in the foreground, become independent, and it is no longer claimed that they still exemplify the theory put forward in Vol. I. In some cases, however, this is certainly still so, and the theoretical analysis of the new case studies is not passed over. Indeed, the notion of reduction, which is based on the theory developed in Vol. I, remains in principle *open*, as indicated, and we are hence free to extend the theory through new types of reduction. According to Vol. I, however, the reductions gained iteratively through a combination of elementary reductions are still linked by the requirement and its validation that they represent possible *empirical progress*. This demand will not be given up in Vol. II, but its specification and supporting evidence will move to the background. We will there consider the more difficult cases (see the announcement of planned chapters) which, in part, regard the questions of reducibility and empirical progress (in the usual sense) that still remain controversial to this day. These cases also became the starting point of attacks against the orthodox theories of the philosophy of science of logical empiricism and critical rationalism of the period from the 30s to the 50s. Under the common battle cry of 'incommensurability', Feyerabend and Th. Kuhn above all carried out such attacks. The notion of incommensurability was to emphasize the difficulties into which one falls in the attempt to reduce, for example, the Newtonian–Galilean space-time theory to the Einsteinian–Minkowskian, or Newtonian gravitational theory to general relativity, or classical mechanics to quantum mechanics. Such attempts suggested themselves for the following reason. Each second-mentioned, later of these theory pairs should be the *successor* of the first-mentioned, earlier theory. However, in the application to overlapping areas they didn't simply disagree, but rather an unpleasant contradiction appeared which according to the Kuhn–Feyerabend analysis should be a *semantic* incommensurability. It remains to be tested whether this analysis is the appropriate way to describe the situation, where the hardest test may be that of the relationship of quantum mechanics to its classical predecessors. The quantum field theories, in so far as they have classic predecessors (such as, for example, quantum electrodynamics has), are also included in this examination. It is certain that the orthodoxy has taken it too easily in all of these cases (and others).

The successor relationship between physical theories referred to above is perhaps the most important historical context for the description of the systematic notion of reduction used—unproblematically for the examples of Vol. I, but with some considerable difficulty for the examples in Vol. II. In addition to those already mentioned, the problematic cases also include the well-known type of *ontological reduction* (or micro-reduction). On the basis of the usual analysis, the reduction here is due to the objects of the reduced theory *consisting* of the objects of the reducing theory in some sense. So this is just half of the basic idea of the classical atomism, to which the claim adds that the analysis into components cannot continue indefinitely but must terminate after finitely many steps. One calls that of which something is made its parts, so the story ends just with the *indivisible* constituent parts of matter. The temporal relationship is, in this case, not so obviously captured because of the long and repeatedly stuck atomistic tradition. In addition, here we deal with a whole chain of reductions (according to the usual analysis), pervading a hierarchy of theories endowed with ever-changing responsibilities. This hierarchy is, according to the mentioned basic idea of this type of reduction, not usually given by the theories themselves, but through their objects. Today, it most often begins with the elementary particles and continues (depending on the degree of differentiation) through the layers of nuclei, atoms, molecules, cells, etc., up to (taking an extreme) human society. This hierarchy extends far beyond the scope of physics. It will concern us only in the physically relevant layers where, above all, (phenomenological and statistical) thermodynamics, continuum mechanics, and solid-state physics are treated. Here, too, the extent to which the accepted philosophical, reductionist notions of that layer structure of matter and of atomism are justified has to be examined here, as well as other anti-reductionist notions of wholeness and emergence.

So much for the preliminary clarification of the character of this book concerning its content. There remains a word to be said about the *rational reconstruction* of physics, which lays the grounds for the substantive reduction-theoretical explanations. Since the early days of logical empiricism, the rational reconstruction of an empirical science has been understood to be the in-depth, as it were, logically purified version of the original presentation given by the respective scientists themselves, achieved through extensive logico-conceptual analysis. In the present case, the substantive ideas could be presented if needed without a special reconstruction, but only out of necessity, since the task is to establish a relatively independent area not cultivated by the scientists themselves. The reconstruction could also be, even without necessity here, differently chosen than is actually done without endangering the reduction-theoretical content. These freedoms are already mentioned at the beginning, so that the degree of detail that is dedicated to the matter, despite everything, does not lead to overestimating the appearance and the special role of the chosen reconstruction. Experts initially vastly exaggerated the benefits of reconstructionism, and later those of different expertise exaggerated its disadvantages. These observations are made in the main text (1.3). The reader familiar with the practices of formal philosophy of science will anyway immediately notice that our reconstruction (like the reduction theory itself based upon it) is merely *sketched*. This should be stated in consolation to those with no familiarity with it.

In keeping with the main subject of the book, the reduction of physical theories, the reconstruction undertaken here concerns mainly the concept of a physical theory. It ultimately fulfills the ages-old claim that a science closed in content should be presented *more geometrico*, that is, given *axiomatically*. But, today, slightly more lies behind this requirement than at the time of its first survey and practice by Aristotle and Euclid. Today, it is required that a logical framework be given first for a complete axiomatization. As such, we choose here a standard first-order *set theory* with individuals. Admitting individuals (in the sense of Zermelo's 'Urelemente', i.e., objects that do not contain elements but nonetheless are not the empty set[1]) is convenient if one wants an approach that extends beyond mathematics. A set theory furthermore equipped with the usual axioms is stronger than any finite-type logic and has the advantage of uniformity in any desired application. However, it primarily recommends itself for coping with the usually high mathematical demands of physical theories. That is, in principle, in the propositional forms of the underlying language of physics, mathematical sets add as referents to physical quantities and urelements. A special case is widely known from model theory, in which the arguments of a propositional form generate a *structure*. The connection with physics is made by reconstructing its main object—the *physical system*—as one structure. To the structure, a proposition (form) is added which refers to the structure (i.e., the physical system). As an axiom of a particular theory, this proposition (form) has to be satisfied (among other, less central claims) by the systems which this very theory qualifies as physically possible. Examples are Maxwell's equations of electrodynamics or the Schrödinger equation of quantum mechanics.

For the use of the first volume of this two-part work lying before you, there is no further advice other than to read it through from cover to cover. To do so is still no golden road to the acquisition of the contents. But, by virtue of the systematic character of Vol. I, in particular, this is the quickest way to arrive at the goal. In Chap. 1, the problem will initially be informally explained: first, how the physicists saw it; then, from the perspective of philosophers of science; and, finally, how it shall be addressed here. In so far as historical material is also laid out in Chap. 1, it appears in most places not for the purposes of dispute, but rather to inform, at least so far as Vol. I is concerned. I consider the theory I present to be more powerful than others with which I have become acquainted (for an exception, see Ludwig 1990 and Scheibe 1987un), but I do not provide a proof of it. Regarding this, the reader must form his own opinion. The systematic treatment of the subject begins with 1.3, where the principles of the theory are presented. Chapter 2 serves entirely to explain the further applied concept of a physical theory. In Chap. 3, a theory of confirmation (for physical theories) is outlined. It is then used in Sect. 3.3 to form a, so-to-speak, degenerate concept of reduction, according to which a theory appears already reduced to another if, roughly speaking, its empirical successes are also always successes of this other theory. It is expected that *any* (genuine) reduction implies this empirical reduction, and this will always be examined. (In this respect, here, too, an adequacy condition in the sense of the Carnapian explanation method is employed.)

The next two chapter titles indicate what is probably the most important division of the elementary reductions: the exact versus the approximate. The neglect of approximate reductions (and explanations) was a major shortcoming of the early efforts of logical empiricism regarding the matter. Even the reduction of the Galilean law to Newton's gravitational theory only succeeds approximately. Exact reductions primarily play a role in compositions, but composite reductions are already approximate if only one of their components is. In all other respects, in the further

[1] Zermelo (1980). Scheibe identifies them with individuals in the sense of Field (1980, 9). See Notes 2 and 3 of Chap. 2 and Sect. 2.2.

subdivisions of Chaps. 4 and 5, we find a quite colorful landscape before us, bearing in mind that the combined reductions still do not yet appear at all there. However, they are just found most frequently in the applications: The classical formula for the mean energy of a harmonic oscillator in thermal equilibrium is (partially) traced back in Sect. 6.2 to quantum mechanics by six successive reductions of three different kinds.

In Chap. 6, the last of the first part, we leave the domain of reductions proper with the introduction of so-called partial reductions. Their introduction is partly a concession to a common practice of physics. It is a frequently found expression that, for example, classical mechanics has proven to be a limiting case of quantum mechanics and Newtonian gravitational theory as a limiting case of the Einsteinian, or the like. In fact, however, this normal textbook manner of speech will never be justified except through the demonstration of more or less special *deductions* (in a broad sense) from classical mechanics, respectively, of Newtonian theory, as such limiting cases. With our 'difficult cases', which we kept especially for Vol. II, one is even happy already having such partial results. In addition—and this is only a concession to human cognition—in the development of physics, one theory could be replaced by another and truly improved also *without* entirely reducing the former to the latter. The partial reductions still possible in this case, thus, form a natural transition to the second part dedicated to the problematic cases, which we will face there again.

Naturally, the book is written for those who have an interest in its subject: the unity of physics. It can be read by anyone who has studied physics and possesses mathematical knowledge and has a sense for logical order as well. Philosophical knowledge is not required. Although the issue is ultimately a philosophical one, and the very questions regarding the unity of knowledge reach back to the beginnings of Greek philosophy, the present contribution is, nonetheless, a less philosophical one and a much more physics-oriented instance of conceptual analysis. From the philosophical side, C. F. von Weizsäcker, in particular, has advocated the unity of physics in many works—most notably in his book *Die Einheit der Natur* (Weizsäcker 1971). Although I readily admit that the unity of knowledge is a worthy ideal to envision, I don't argue for it in this book—at least not directly and primarily. Instead, it's more about answering the question of what someone who feels himself reductionist would have to do *in concreto*, if he wanted to give his feeling general validity, and if he is ready to submit to certain now usual, but not too immodest standards of philosophy of science. There is, here, an intermediate realm between the physical and philosophical issues in the narrow sense, and just questions about reduction and explanation have their place here. In his not yet too old book *Dreams of a Final Theory*, Steven Weinberg, the currently most prominent physicist–spokesman for reductionism, expressed the view "that a knowledge of philosophy does not seem to be of use to physicists" (Weinberg 1992, 168). One can only agree. The esotericism of philosophy consists in the fact that it only brings benefits to the one that already comes to it with a philosophical question. When Weinberg then continues on "—always with the exception that the work of some philosophers helps us to avoid the errors of other philosophers", one can only hope that this help is only a minimum requirement. Weinberg himself is already within that intermediate realm where physics and philosophy merge into one each other when, with respect to the standard model of elementary particle physics, he says "and now we want to take the next step and explain the standard model [...] I do not understand how this can not seem to be an important task to anyone who is curious about why the world is the way it is [...]" (58). Can—one wonders—a philosopher do no more than help the physicist avoid the mistakes of other philosophers?

References

Field, H. 1980. *Science without Numbers. A Defence of Nominalism*. Princeton, NJ: Princeton University Press.

Ludwig, G. 1990. Die Grundstrukturen einer physikalischen Theorie. Berlin: Springer.

Scheibe, E. 1987un. Explanation, Reduction, Progress. Investigations into the Unity of Physics. A Seminar given at the Philosophical Department of the University of California, Irvine 1987. Unpublished lecture. https://doi.org/10.17877/DE290R-22744

Weinberg, St. 1992. *Dreams of a Final Theory*. New York: Vintage Books.

Weizsäcker, C. F. v. 1971. *Die Einheit der Natur*. München: Hanser. Engl.: *The Unity of Nature*. New York: Farrar Straus Giroux 1980.

Zermelo, E. 1908. Untersuchungen über die Grundlagen der Mengenlehre. *Mathematische Annalen* 65, 261–281.

Contents

Chapter 1
The Problem

This book raises a systematic claim. It involves the presentation of a theory of the reduction of physical theories to others. In connection therewith, it regards the questions of the theoretical progress of physics and its development towards a unity. These questions are closely bound up with the problem of reduction: A theory that can reduce or explain another theory is expected at least not to be worse and, usually, even better than its predecessor. It represents a step forward. And if we can reduce several theories to one and the same theory, we have achieved a unification of the former. Insofar as this is the case and generally newer theories explain or reduce their predecessors, it is about the development of physics and, thereby, the history of physics as well. But not in the way a historian sees it. In the foreground stands the question: What do we mean by theory reduction, what do we mean by the progress and unity of physics so that these concepts help to make broadly comprehensible the course of physics with hindsight? All historical details, the detours that physics has taken, external influences on its developments, the personalities and the interactions of eminent physicists, etc.—all of that is not considered here. Instead, the book is about the conceptual reconstruction of physical theories and inter-theoretic relations. In addition, it is systematic in the other sense that it does not reproduce the history of the concepts of reduction, progress, and unity as others have developed them (within and outside of physics) and is rather a self-contained presentation of the matter itself. The citations and allusions to other relevant work mentioned, although in total numerous, are not given for the purpose of supporting what is said in this book. They serve much more as indicators of sources and supplements to what is said.

A certain exception to this is made by the first chapter. It seems reasonable, as an introduction to an otherwise systematically laid out book, at least to point out the previous fate of the addressed issues, in order to make the tradition clear to which the further exposition belongs. Here, the peculiarity becomes immediately apparent that, historically, one has to distinguish two traditions that were only loosely connected with each other, even though they struggled with the same problem. On

E. Scheibe, *The Reduction of Physical Theories*, Fundamental Theories of Physics 207,
https://doi.org/10.1007/978-3-662-65000-4_1

the one hand, we have the efforts of physicists of understanding the progress of their science (Sect. 1.1). On the other hand, the philosophy of science of our century directed its conceptions, above all, toward physics and arrived, despite the absence of closer contact with the physicists, at somewhat similar and, in particular, similarly controversial conclusions as the physicists themselves in just the conceptual field of interest to us, that of reduction, progress, and unity (Sect. 1.2).

The truly controversial point is, in short, the question of how far we can push reductionism even within physics, not to mention such grandiose perspectives as the reduction of chemistry to physics or of biology or even psychology to chemistry and, thus, possibly also to physics. In other words, it regards the question of how unified our physics is in its present stage, at which it remains still composed of many theories and there exists no final clarity regarding their independence of or dependence on these theories. From the part of the history and philosophy of science, Kuhn and Feyerabend have drawn a gloomy picture, and their respective thoughts were anticipated by some physicists, in particular, by Bohr and Heisenberg. The following book starts from this situation and attempts to cope with it by a moderately reductionist tendency. Even an only nearly complete examination of physics with reductionist intentions would, however, be an undertaking which can only be described as a hybrid, especially if one takes into account that a large portion of the work must be paid to the question of which reduction concept and which notions of progress and unity one wishes to employ. Therefore, the only undertaking that currently makes sense is one that places the systematic development of a concept of reduction in the foreground (Sect. 1.3) and explains it through as many and varied examples as possible, which the abundance of theories of modern physics provides.

This is done in this book, and the underlying idea on which it is based is *new*, given that most of the recent larger attempts to cope with the situation created by Kuhn and Feyerabend undertaken in the philosophy of science have proven themselves not very helpful for the development of our own thoughts. As deserving as these works might be in one way or another, an engagement with them at this point is omitted (Stegmüller 1976; Krajewski 1977; Yoshida 1977; Spector 1978; Dilworth 1981; Niiniluoto 1984; Balzer et al. 1987). It is different in the case of the theory set out in (Ludwig 1990), by which the present study was strongly influenced but has, nonetheless, finally moved beyond (cf. especially the Appendix).[1]

1.1 The Physicists' Conception of Progress

The rapid and, in some respects, unusual development of physics since the mid-19th Century did not lead only to occasional comments of physicists about the peculiar way in which their discipline developed. Rather, one can speak of the gradual emergence

[1] Scheibe planned to add to Vol. II an appendix about Ludwig's approach, which however was finally not realized; see Table of Contents Vol. II at the end of this book. Scheibe (1982a) compares the approaches of Ludwig (1990) and Sneed (1971). In an unpublished lecture, Scheibe (1987un, Ch. VIII) explained Ludwig's approach in more detail.

of a proper tradition of thinking about the theoretical progress of physics. It can further be seen that, in this tradition, largely uninfluenced and unnoticed by philosophers, many important ideas that we recall from the recent philosophical controversy about the matter are anticipated in principle, i.e., with the degree of detail that one can reasonably expect of physicists. In this section, we will want next to provide an historical picture of this tradition (see Scheibe 1988a), which will only in outline be connected to the recent philosophical discussion.

The basic features of the concepts of progress of physics developed within the community of physicists were already expressed in mature form at the end of the last century. Even at that time—1895— *Boltzmann* made the following remarks in an obituary of Joseph Stefan (Boltzmann 1905, 94–95) (see also (Boltzmann 1979, 59–60, 123–124 and 207–208))—to begin with, I quote only the first part of the passage:

> The layman may have the idea that to the existing basic notions and basic causes of the phenomena new notions and causes are gradually added and that, in this way, our knowledge of nature undergoes a continuous development. This view, however, is erroneous, and the development of theoretical physics has always been one by leaps. In many cases, it took decades or even more than a century to articulate fully a theory such that a clear picture of a certain class of phenomena was accomplished. But, finally, new phenomena became known which were incompatible with the theory; the attempt to assimilate the former into the latter was in vain. A struggle began between the adherents of the theory and the advocates of an entirely new conception until, eventually, the latter was generally accepted.

It is fairly obvious that the course of development of theoretical physics here outlined by Boltzmann broadly resembles the more general model of scientific development which recently Thomas Kuhn (1970a) proposed again and carried out in more detail. After a phase of continuous development (Kuhn's normal science), the discipline runs into problems which it initially tries to master within the framework of the prevailing doctrine (Kuhn's crisis). Finally, however, there comes "an entirely new conception" which supersedes after some "struggle" (Kuhn's scientific revolution). It is here where, according to Boltzmann's text, the development makes a "leap."

But this is only half of Boltzmann's (and also Kuhn's) story. In the second part, Boltzmann specifies this discontinuity as follows:

> Formerly one used to say that the old view has been recognized as false. This sounds as if the new ideas were absolutely true and, on the other hand, the old (being false) had been entirely useless. Nowadays, to avoid confusion in this respect, one is content to say: The new way of ideas is a better, a more complete and a more adequate description of the facts. Thereby it is clearly expressed (1) that the earlier theory, too, had been useful because it gave an, if only partially, true picture of the facts, and (2) that the possibility is not excluded that the new theory in turn will be superseded by a more suitable one.

The discontinuous development of theoretical physics, which Boltzmann emphasized in principle, is thus explicitly mitigated here, and the remark that well-proven and reliable physical theories will never be completely wrong is one of the most repeated (and now largely unconsciously so) in the physics literature. Kuhn also put the scientific-revolution surviving and progressive elements together, but on the whole put more stress on the incommensurability of the new conception with the old. As we shall see, the question of how large each jump is has been answered in different ways by physicists. Either way, Boltzmann's view, if you take his words somewhat with a pinch of salt, is still that believed by physicists.

A first refinement we owe to Nernst (1911, 4–5), for introducing the concept of *the limit of applicability*. He thereby stresses the conservative component of the Boltzmannian schema: Of course, many well-proven reliable laws must ultimately have been revised. However:

> If we consider the matter more closely, it is obvious that the law in question has retained its validity over a wide range, but that the *limits of its applicability* have been more sharply defined. It can even be said that since the development of the exact natural sciences, there is scarcely one law established by an investigator of the highest rank which ...has not remained a serviceable law of nature *within certain limits*.

The refinement of Fresnel optics by Maxwell's electrodynamics is then given as an example.

On the other hand, somewhat later, instructed by what occurred with general relativity, Nernst again stressed the revolutionary component of Boltzmann's notion of development. In his Berlin rector's address of 1921, in which Nernst explicitly refers to Boltzmann, he explains with Einstein and Newton's theory in mind (Nernst 1922, 489):

> The modifications that have to be made with respect to the earlier theory are so small that, according to the present state of research, they can be neglected, except for the computation of the orbit of Mercury. But, as a matter of principle, every computation that astronomers have performed so far must be changed. And it is this principal aspect of the problem, not the numerical amount of the correction, that is our point.

This passage shows that the physicists are not all quite as unscrupulous as Feyerabend would like to see them when he says that they do not check, upon the appearance of a new theory, whether it permits the reproduction of the empirical successes of its predecessor—indeed, that they may not even feel the obligation to do so (Feyerabend 1970, 276–7). Nernst, however, feels that obligation when he recalls here that all calulations based on Newton's theory actually had to be re-created ab initio on the basis of Einstein's theory. Despite large computers, such undertakings would of course even not be begun today. That, however, *some* general insight must be won, which guarantees that if you followed all those new calculations then the Einsteinian results would not lag behind the Newtonian ones, will surely be the common conviction of all physicists who speak of progress in physics. It must be admitted that the general reflective analyses which offer such great assurance have not yet been completed because of the great mathematical difficulties involved (cf. Chap. 2 of Vol. II).

Nernst (1922, 491–2) takes his cause still further and—already fully in the spirit of the later programs of Kuhn and Feyerabend—makes clear how deep the break between old and new theory can be:

> One might think that ...the laws of nature ...are valid with absolute precision in certain domains and that the matter could be settled very easily by pointing out the limits within which they remain valid. For all practical applications this is true enough ...From a strictly logical

> point of view, however, the matter appears much more disastrous. If a general law of nature becomes significantly inaccurate beyond certain limits, then the curse of this imprecision comes to roost on every application of the law even within these limits, though the magnitudes of the errors are below the threshold of measurement for the time being.

The contradiction between the old and new theory, because that is the point here, is consequently not so limited that it vanishes exactly outside these limits. Even if we could increase the accuracy indefinitely, we would have to recalculate always and everywhere with the new theory. The old theory is at best an approximate, *limiting case* of the new one. (The notion of the limiting case was already used in this sense by Hertz (1892, 26–28), and soon after also by Einstein (1914, 740–742, and 1917, 50), for this time also cf. Hilbert (1919, 52–55)).

Today, the limiting case scenario already finds itself presented at the textbook level. In their book about gravity, Misner et al. (1973, Sect. 17.4) develop the view that

> as physics develops and expands, its unity is maintained by a network of correspondence principles through which simpler theories maintain their vitality by links to more sophisticated but more accurate ones.

As examples, the authors mention geometrical optics, Newtonian mechanics, thermodynamics, and Hamiltonian mechanics as so-called 'correspondence principle limits', that is, as limiting cases in the sense of correspondence principles, to physical optics, relativistic mechanics, statistical mechanics, and quantum mechanics, respectively. They then study more precisely the correspondence structure of the general theory of relativity, and indicate at least four limiting cases, of which one is Newton's gravitational theory. "In all these examples and others," the authors summarize,

> the newer, more sophisticated theory is "better" than its predecessor because it gives a good description of a more extended domain of physics, or a more accurate description of the same domain, or both.

There is, however, not only this empirical superiority:

> The correspondence between the newer theory and its predecessor gives one the power *to recover the older theory* [and] can be exhibited *by straightforward mathematics* (emphasis mine)

Up to this point, one could say, the talk has been predominantly of *local* progress—of the question of that of which the direct advance from one theory to the next consists. But physicists have themselves also come up with ideas as to 'how the story continues'. We want now to investigate the more *global* aspect of progress. First, there is the aspect of the *transitivity* of the previously described progress: If T_1 is a step forward compared to T_2 and T_2 compared to T_3, so is T_1 compared to T_3. For what we usually mean by a step forward, this condition appears self-evident. But it is not so for that by means of which we capture the content of progress, e.g., a borderline case, a reduction, etc., and it is even less so for the actual development of physics as a whole. But, if we examine the transitivity to be fulfilled in content as well (and physicists do), there arises here a first possibility to extend our recent insights to *greater periods* and the corresponding greater content of theories.

Such a thing also even happens sometimes today in textbooks. Guillemin and Sternberg (1984, 4) initially present themselves as supporters of Boltzmann's conception of theory development, as can be seen when they say,

> In the history of physics it is often the case that, when an older theory is superseded by a newer one, the older theory still retains its validity—either as an approximation to the newer theory, an approximation that is valid for an interesting range of circumstances, or as a special case of the newer theory.

And then they give the following overview of the development of optics from Gauss to the present, with the special case indicated by the plain arrow and the double arrow meaning the case of the limiting approximation:

> Gaussian optics $\rightarrow$ linear optics $\Rightarrow$ geometrical optics $\Rightarrow$ wave optics $\rightarrow$ Maxwellian Electrodynamics $\Rightarrow$ quantum electrodynamics

It is entirely in the spirit of such multiple-kind (here with two kinds) and iterative theory reductions that we will build the programme of our way forward in Sect. 1.3.

In connection with transitivity, one can still go one step further without having to abandon the consensus among physicists. The developmental step to greater universality is often also accompanied by a *merging together* of previously apparently independent theories into a more comprehensive theory. The linear, transitive hierarchy thereby takes on a tree-like structure. Kepler's laws and Galileo's law of falling bodies became united in Newton's gravitational theory. The long separately developed theories of electric, magnetic, and optical phenomena were combined in Maxwell's electrodynamics. That this always has provided particularly great successes and that such successes are also desirable in principle is broadly agreed. So we hear from *Planck* in his famous Leiden lecture of 1908 (Planck 1949, 31):

> the signature of the entire previous development of theoretical physics is a unification of its system ....

This Planck (1949, 28–29) immediately explicates:

> In physics as an empirical science ... it has occurred frequently and still also happens that two theories that have achieved a certain independence, in their further development collide with each other and must be modified in order to remain compatible. In this mutual adaptation of different theories lies the main seed of their fruition and later development into a higher unity.

Very recently, *Steven Weinberg* has used with refreshing abandon stronger, philosophically loaded terms, such as explanation, reduction, and unity, to give expression to his fundamentally reductionist attitude. For Weinberg (1987, 435), "there is a sense of direction in science, that some generalizations are 'explained' by others". Such explanations of theories have two important properties. For one, they are transitive (see Mayr and Weinberg 1988, 475):

> If a large number of facts $a, b, c, \ldots$ are explained by a set of theories $X, Y, \ldots$, and then these theories are in turn explained by a more satisfying theory Z, then I would say that the facts $a, b, c, \ldots$ are explained by theory Z ....

For example, the development of general relativity was important "because it explained the theories of Newton that had earlier explained so much else." Second, we also have here some convergence, the equal binding together of the branched network of physical theory in a certain direction (Weinberg 1987, 435):

> There are arrows of scientific explanation that thread through the space of all scientific generalizations. Having discovered many of these arrows, we can now look at the pattern that has emerged, and we notice a remarkable thing: perhaps the greatest scientific discovery of all. These arrows seem to converge to a common source! Start anywhere in science and, like an unpleasant child, keep asking 'Why?'. You will eventually get down to the level of the very small.

Considering here the idea of historical global progress (in the specific sense of the physicists), although it goes beyond local progress, does not yet mean that we are dealing already with the idea of a development towards *the unity of physics*. This has been emphasized, for example, by David Bohm. Although this runs entirely in the tradition begun by Boltzmann and, in that respect, includes even successor relationships as difficult as that between classical mechanics and quantum mechanics, Bohm (1980, 4–5) has explicitly noted:

> In this activity, there is evidently no reason to suppose that there is or will be a final form of insight (corresponding to absolute truth) or even a steady series of approximations to this. Rather, in the nature of the case, one may expect the unending development of new forms of insight (which will, however, assimilate certain key features of the older forms as simplifications, in the way that relativity theory does with Newtonian theory).

(Similarly, also Bondi 1977.) Others have been more optimistic regarding this point and have, in one sense or another, expressed the unity of physics as an objective and a hope. The successes of quantum mechanics and the further development of nuclear and particle physics have encouraged many physicists in this direction.

Planck (1949, 106) had already said as much in a lecture of 1915:

> ...The main goal of any science is and remains the fusion of all their grand theories into a single one, in which all the problems of science find their unambiguous place and their unambiguous solution.

In addition, there is for Planck "only a single science, and this is obligatory for all mankind, and it does not waver, but rather it strides forward." But to this integral perspective Planck adds, nonetheless, "even if it will never achieve the ideal goal and can never reach it." (Quoted in Kangro 1970, 227.)

There was also optimism in the generation of the 'young men' who then created quantum mechanics. In the late twenties, Dirac (1929, 714) begins an article with the remark that

> The general theory of quantum mechanics is now almost complete...The underlying physical laws necessary for the mathematical theory of a large part of physics and the whole of chemistry are thus completely known ...

And, as late as 1966 in Dirac's lectures on quantum field theory, it is said—now somewhat more cautiously (Dirac 1966, 1–2):

> Our object is to get a single comprehensive theory that will describe the whole of physics ...I do not need to tell you that such a theory has not yet been attained. It is the ultimate goal toward which all physicists are working.

These statements are not necessarily at odds with the traditional view that in the development of physics there appear significant jumps. Because the inclusion of older theories in their successors can, anyway, only be maintained to the extent that they are empirically established. Larger jumps are simply a reflection of errors committed during the establishment of older theories, and errors do not stand in the way of the unity of physics *in so far as they are corrected.* Big jumps are great progress because they eliminate large errors.

The conviction that the closure of physics lies in a theory of elementary particles and can not be long in coming is also shared today by von Weizsäcker and Hawking. From the former, we hear (Weizsäcker 1980, 154):

> Assume the theory of elementary particles, including also the theory of gravitation, has been completed—and couldn't this easily be the case still in our century? Then, at least in the domain we today call physics, no special law of nature whatever would be left—special in the sense of a proposition theoretically underivable, in principle, from the basic law.

Hawking (1980, 1–2) expressed similar thoughts in his inaugural lecture 'Is the end in sight for theoretical physics?'. He considers it possible "that the goal of theoretical physics might be achieved in the not too distant future ...". He described the goal as "a complete, consistent and unified theory of physical interactions which would describe all physical observations." And if Hawking warns one "to be very cautious about making such predictions," he sees "some grounds for cautious optimism that we may see a complete theory within the lifetime of some of those present here."

To this day there is no such theory. Yet we find the pros and cons of the unity of physics strenghtened by proposals for a characteristic mark of unified or definite theories, and there is the idea of theories beyond which no normal further development is possible, and these are still not yet the fundamental theory of physics. Seemingly prophetic utterances stem from a time when there did not even exist a rudimentary theory of elementary particles. But with a view to the modern development of physics, and based on a wholeness criterion going back to Aristotle (cf. Scheibe 1987a), Planck in the aforementioned lecture of 1908 compared previous physics with a collection of paintings from which from you could remove each painting "without affecting the others. It will not, in that case, be possible in the future physical world-picture. No single feature thereof can be omitted as insignificant ..." (Planck 1949, 45–46).

Already in 1919, the time had come for *Einstein* to draw upon a similar thought regarding a theory already at hand, namely, his own theory of gravitation. "The great attraction of the theory"—ventures Einstein (1919, 43)— "is its logical consistency. If any deduction from it should prove untenable, it must be given up. *A modification of it seems impossible without destruction of the whole.*"[2] And, thirty years later,

[2] Emphasis by E. Scheibe.

Einstein (1950, 15) judges his so-called unified field theory: "In favor of this theory are …its logical simplicity and its 'rigidity.' Rigidity here means that the theory is either true or false, *but not modifiable*" (emphasis mine). This is, of course, not a simple application of the Popperian falsification criterion or the *tertium non datur.* The view is that a falsification test speaks not only to *part* of the theory, but that the theory is either true in its entirety or *entirely* false, respectively. In addition, Einstein would not have undertaken his intense but ultimately just as fruitless efforts to find a unified theory of gravitation and electromagnetism if he had held general relativity theory to be definitive. Regarding the end of physics, he was in doubt.

> We do not know whether or not this ambition will ever result in a definite system. If one is asked for his opinion, he is inclined to answer no. While wrestling with the problems, however, one will never give up the hope that this greatest of all aims can really be attained to a very high degree. (Einstein 1936, 353)

The question as to how Einstein imagined the relation of a possible final theory to general relativity if *both* shall have that characteristic of rigidity remains to be asked.

Apparently independently, physicists involved mainly in the development of *quantum theory* as a truly revolutionary innovation compared with classical physics (including the special theory of relativity) arrived at conceptual pictures entirely similar to those of Einstein. The underlying difficulties here in building a bridge, especially with quantum mechanics viewed as an improvement over classical mechanics, were first sensed by *Bohr*. Already for Bohr's atomic model, it had been recognized that in the limit of large quantum numbers the (mechanical) angular frequencies of the electron well approximate the (mean quantum) radiation frequencies. In hindsight, knowing the solution to the problem in quantum mechanics, Bohr (1934, 85) says, however:

> The [asymptotic connection of atomic properties with classical electrodynamics, demanded by the correspondence principle] means that in the limit of large quantum numbers, where the relative difference between adjacent stationary states vanishes asymptotically, mechanical pictures of electronic motion may be rationally used. It must be emphasized, however, that this connection cannot be regarded as a gradual transition towards classical theory in the sense that the quantum postulate would lose its significance for high quantum numbers.

Moreover, Bohr's attempts at a general formulation of the relationship of quantum mechanics to classical physics via his *correspondence principle* always contain the warning that this principle, despite the apparent limiting cases, connects two fundamentally different theories. With this principle, one aims at "regarding quantum theory as a rational generalization of the classical theories" (ibid., 70). With this, one has—again reminiscent of Boltzmann—first, "the demand of a direct concurrence of the quantum-mechanical description with the customary description in the border region where the quantum of action can be neglected" (ibid., 100). On the other hand, it is about "a reinterpretation which fulfils this demand *without* being at variance with the postulate of the indivisibility of the quantum of action" (ibid., 110).[3] The most detailed considerations of a general nature regarding our present problem, on the

[3] Emphasis by E. Scheibe.

part of physicists, come from *Heisenberg* (see the overview in Scheibe 1993a). They are characterized by his concept of *closed theories*. Heisenberg (1971, 96) called a theory closed if "inasmuch as we can describe a particular phenomenon with the concepts [of this theory], the laws [of this theory] hold quite rigorously [...]." This notion of closedness seems, at first glance, to capture something quite different from Einstein's notion. According to a consideration of von Weizsäcker, this is nonetheless not the case. He called a theory closed if it cannot be corrected by means of small changes (Weizsäcker 1980, 156). This is negligibly different from Einstein's notion of rigidity. The question is only that of what *small* and *big* changes are in these notions. Now, the most conspicuous elements of a theory are its concepts and laws. It therefore makes sense to talk of small changes where one only changes the laws, e.g., by a corrective term, but of of large changes, if the conceptual structure of the theory is thereby already modified. With this understanding, the two criteria are then equivalent.

The questionable connection between concepts and laws of a theory was seen early on by Pauli when he wrote in a letter of Sept. 20, 1923 to Eddington (Pauli 1979, 116)[4]:

> I might emphasise especially that the quantum theory in no way calls for only a modification of light theory, but generally calls for a new definition of the concept of electromagnetic field for non-static processes. Indeed the classical theory fails also in the description of electron-atom collisions, as well as for the behaviour of the electron in the atom in time-varying fields (dispersion theory). The famous contradiction between the interference capability of light emitted in various directions and its property of remaining constantly disposed in energy quanta (Einstein's "light-quanta"), for example in the photoelectric effect, comes only from the fact that we give up the laws of classical theory, but still always work with the concepts of that theory.

Heisenberg did explicitly not develop his concept of the closed theory for a characterization of the 'final' theory. He hardly believed in its possibility (cf. Heisenberg 1974, 39–46). But, he recognized the exceptional nature of some theories *that we already have*, such as Newtonian mechanics, classical electrodynamics, and quantum mechanics, in that these theories cannot be further developed in the usual sense. We cannot *merely* extend their scope, we cannot *merely* modify their laws—the only change yet to come into play is the change of its concepts so as actually to *suspend* the theory. Even a union of two such closed theories in a higher unity is inconceivable and no progress can eliminate them:

> The edifice of exact science can hardly be looked upon as a consistent and coherent unit in the naive way we had hoped. ...for it consists of specific parts, and though each of these is connected to the others by many passageways and each may encompass some others or be encompassed by others, nevertheless each is a unit complete in itself. (Heisenberg 1952, 27)

While Heisenberg and likely also Einstein wanted to apply their conceptions of a closed theory and a rigid theory, respectively, to theories that are not already the final theory of physics, a similar concept was recently used by Weinberg in direct connection with the question of the final theory. With Einstein's terminology, Weinberg

[4] Translation taken from (Hendry 1984, 20).

(1992, 17) suggested "to identify the final theory as one that is so *rigid* that it cannot be warped into some slightly different theory without introducing logical absurdities like infinite energies". Because we do not have the final theory, Weinberg must test his concept on other theories, and quantum mechanics is to him the most promising candidate. He reports his specifically undertaken trial attempt to alter quantum mechanics with a slight non-linear correction (ibid. pp. 85–89). This attempt failed, and Weinberg suggested therefore not only that quantum mechanics meets his criterion, but even that "quantum mechanics may survive not merely as an approximation to a deeper truth …but as a precisely valid feature of the final theory (ibid. 89).

Although this assumption would be a way out, one would nonetheless really prefer not to believe that a rigid theory could contain even such a rigid part. And one actually finds Weinberg, without his explicitly noting it, with a *comparative* concept of rigidity of a theory, i.e., a notion that one theory is more rigid than another (ibid. 104–107). For example, Einstein's theory is more rigid than Newton's, as easily modified characteristics of the latter are themselves recovered as not easily modified features of the former. Newton's law of gravity is almost arbitrarily modifiable without even affecting the foundations of mechanics; similarly for the equality of inertial and gravitational masses. In Einstein's theory these assumptions are, by contrast, compelling consequences following from very basic assumptions (and constraints) of a geometric nature. There exists no geometric theory of gravitation that leads to other than the $1/r^2$-dependence that does not appear *substantially* different from Einstein's.

The general considerations of physicists, stimulated by the appearance of the general theory of relativity and quantum theory, thus show that the unanimity regarding the theoretical progress of physics which is easily reached in less difficult cases is lost in the more difficult cases and for more difficult questions, and, therefore, the concepts of progress, unity, and unification of physics become fundamentally problematic. These considerations show yet again that physicists, in their reactions to these recent theories, anticipated later scientific philosophical considerations, such as those set out by Popper, Kuhn, Feyerabend, Lakatos, and others (see Sect. 1.2 and Lakatos and Musgrave 1970). For, even these philosophers and historians of science, with far-reaching innovations in a scientific discipline in mind, have tried to make clear in such cases the difficulties of describing the development as progress, speaking of approximation to truth, etc. A striking difference between the mutual considerations lies in the fact that philosophers of science try more to describe the developmental *relations* in which successive theories stand to each other, while physicists have attributed certain *characteristics* to theories that put them necessarily in that difficult relationship. As we will see now, the interest of philosophers of science has been directed by logical standards more towards considering the relations between theories than the characteristics of theories—older standards, developed on the basis of relatively simple conditions.

1.2 The Legacy of the Philosophers

One such standard was the traditional notion of *explanation*. That they can only explain but not understand the phenomena had been dictated to the natural sciences by the humanities-oriented philosophy of the late 19th Century. And it was again the logical aspect of explanation—the subsumption of the explanandum under a generality—that was emphasized. In fact, then, the new logical empiricism was concerned about the explication of a scientifically adequate notion of explanation, but initially without this notion being connected with the development and progress of science. This could not happen as long as the interest in explanations of contingent events was in the foreground. The result of that explication, the so-called *DN model* of explanation, was conceived so generally that, in principle, it also allowed laws to serve as explananda. But the creators of the respective notion, *Hempel and Oppenheim*, limited their (full) 1948 explication (though not its adequacy conditions) to the explanation of contingent events because "the precise rational reconstruction of explanation as applied to general regularities presents peculiar problems for which we can offer no solution at present" (Hempel 1965, Sect. 10; here: 273, Note 33). Thus, the explanation of laws was somewhat lost from view and, with it, the possibility of applying the DN model also to the progress of a science: the explanation of the valid parts of obsolete laws by their successors. Hempel only addressed this possibility again when it was already clear that it would never be accomplished with the original DN model but rather with an approximate version thereof (ibid. 343–349).

It is somewhat different with the *concept of reduction* of *Nagel*, developed almost simultaneously. It was concerned from the outset with the reduction of theories to other theories, and the interest in this process was motivated, amongst other things, by the fact that "the phenomenon of a relatively autonomous branch of science becoming absorbed by, or 'reduced' to, some other discipline is an undeniable and recurrent feature of the history of modern science" (Nagel 1949, 99–100 and, again, Nagel 1961, 336–337). The criticism, repeated from the 60s onward, that the new empiricism did not concern itself with the development of science, is therefore not plainly to be raised. Another question, of course, is whether Nagel's reduction concept was adequate and whether it especially came into question to capture theoretical progress. And we first encounter the same difficulty as in the DN model: "In a D-N explanation", says Hempel (1965, 337), "the explanandum is a *logical consequence* of the explanans" (emphasis mine), and we hear similarly from Nagel (1949, 119): "The objective of the reduction is to show that the laws or general principles of the ...science [to be reduced] are simply *logical consequences* of the assumptions of the [reducing] science" (emphasis mine; likewise in Nagel 1961, 352). If one takes this condition seriously, then it is clear that, regardless of what progress may consist of, it can not be the case that a theory is modified in content by its successors. The progress, if any, would be rather cumulative, and even such progress—as we have seen in 1.1—would would then fall outside the scope of the physicists' picture. It should be added that, in this picture, the more general, more comprehensive theory is the reducing theory. The most obvious explication of this idea would be that the reduced theory is not a logical consequence of the reducing, but vice versa (see Sect. 4.1). The central adequacy condition for explanation and reduction therefore comes into conflict with this picture. As mentioned, Hempel later widened his notion through an approximate version, while Nagel worked as if he had always had one in sight. For he stated in a reply to Feyerabend's critique, "the derivation of laws from theories usually involves simplifications and approximations of various kinds, so that even the laws which are allegedly entailed by a theory are in general only approximations to what is strictly entailed by it" (Nagel 1970, 121).

In addition to the (D-) condition of deducibility of the DN model, there is primarily the requirement of the nomological character of explanation and reduction, indicated by the 'N'. When it comes to the case of the explanation of a single event, it tells you that among the explanatory premisses there must be one with *lawful* strength. This does not yet address the difficult question of what it means—what deserves to be called a 'law of physics'. It serves only to emphasize what in the case of explaining an event is clear enough, namely, that due to the N condition, the premisses of an explanation are placed at two very different levels, one lawful and one event-contingent. But what becomes of the N condition if the explanandum, too, is a law-like statement? This question, left open by Hempel and Oppenheim as mentioned, has not been clarified up to the present. It could simply be, however, that really not much has to be changed, since from a combination of law-like and event-contingent premisses just as valid (or almost as valid) lawful conclusions can be drawn as well as event-contingent (perhaps approximate) ones.

With this we come to those premisses which must occur *in addition* to laws in order to enable the deduction (if at all possible) of a given explanandum. In the explanation of events, their occurrence is indispensable because statements that describe single events will never follow from laws alone. Descriptions of other events must add. Hempel (1965, 351) saw this clearly, emphasizing that "the best examples of explanations conforming to the D-N model are based on physical theories of deterministic character". Here, then, the so-called *initial conditions* are the premisses in question. It becomes more difficult again if the explanandum already has the character of law. Laws can indeed follow from other laws, e.g., the conservation laws from an equation of motion of mechanics. It is questionable whether, conversely, laws arise as consequences of other statements *without* these in turn being throughout laws. We see Nagel (1961, 58) wrestle with this problem when he points out, with respect to the question of the status of Kepler's laws with respect to Newton's theory of gravitation, that "it is far from certain that such statements as Kepler's are in fact logically derivable ...from fundamental laws *alone* (as is required ...if these statements are to be classified as laws). There appears to be no way of deducing the Keplerian laws from Newtonian mechanics and gravitational theory ...*without* using additional premises whose predicates are not purely qualitative".

It is interesting to see how Nagel is only half on the right track regarding this issue: Of course, Kepler's laws cannot be obtained as statements about our planetary system without introducing the specific celestial bodies. However, the real problem does not lie here at all. Apart from the fact that for Kepler's laws only an approximate deduction is possible, for them to be directly deducible from Newton's theory *alone* cannot be considered here, because every solution of Newton's equations would then also be a solution of Kepler's laws, and that would be absurd, of course: Progress was made with Newton's theory exactly in that it allowed for completely new sorts of trajectory for gravitating bodies (e.g., the Sun-Earth-Moon system). It is these new physical possibilities that need to be excluded by additional premises (the initial conditions) in order to recover Kepler's laws. And this consideration applies to any system similar to our solar system. As already said, laws can restrict the scope of other laws.

Nagel's (and Hempel's) main problem, however, was not the additional assumptions that came about in this way. His main problem was that of the 'conditions of connectability' in a *heterogeneous reduction*—likewise, additional assumptions as well, but different in origin (Nagel 1961, Chap. 11, and similarly Hempel 1966, Chaps. 5 and 7). Nagel distinguished between homogeneous and heterogeneous reductions according to whether the concepts of the reduced theory (including their scope) are already 'contained' in the conceptual framework of the reducing theory (including its scope) or not. A heterogeneous reduction within physics on which Nagel took guidance is the reduction of the phenomenological thermodynamics to classical mechanics, in particular, the reduction of the concept of temperature to the kinetic theory of gases. This is a case wherein the difficulty of reduction focuses on the question of how one wants to have the *concepts* of the reducing theory, for example, the concept of temperature, explained via the reducing theory, here the kinetic theory—a difficulty which increases when the attempted reduction of chemistry, or even biology, to physics involves explaining chemical or biological concepts, respectively, via the concepts of physics. Nagel saw here not only the difficulty of the *actual determination* of the respective connections in every single case, but also the difficulty of *describing* these 'conditions of connectability' (in Hempel: bridge laws or—more generally—descriptive definitions) *in general*. For this seems to be about relations which are neither analytic relations nor empirical in an ordinary sense. As with Frege's famous example of the identity of the morning star and the evening star, a *semantic* aspect comes into play if we, for example, identify the temperature with the average kinetic energy of the molecules in the (simplified) kinetic theory.

In accordance with the respect Nagel had for this issue, he (wrongly) left the homogeneous reductions behind, and took the reduction of the laws in the heterogeneous case purely deductively, viewing them as routine whenever the 'conditions of connectability' are available as additional premisses. Nagel's distinction, partly under different terminology, has been of continued interest (Sklar 1967; Brittan 1970, D-reduction versus P-reduction; Nickles 1973, domain preserving versus domain combining reduction; Spector 1978, direct theory replacement versus concept replacement; Friedmann 1982). In this book, the distinction will not go unnoticed (see Sect. 4.4), but it will also not be a point of central concern. It belongs to one of those very general distinctions, which are happily encountered in epistemological investigations but we would like, if possible, to avoid it as a topic of explication. The division of homogeneous and heterogeneous reduction is not very clear-cut, if only because the homogeneous reduction of laws may be the presupposition for the heterogeneous formation of concepts in the reducing theory (see Sect. 4.3, conclusion).

Both Hempel and Nagel, of course, made empirical claims regarding their concepts of reduction and explanation. Remarkably, however, these claims concerned only the particular theory or statement units *themselves* and, therefore, no conclusions about whether *empirical progress* was achieved with the explaining or reducing theory compared with the explanandum or the theory to be reduced, respectively. The question of course makes no sense for mere event explanations, but it certainly does for any theory explanation—for example, whether more phenomena are capable of explanation via the explaining theory than via the explained theory. With all respect to relations only between theories—what would be gained for physics by a reduction of theory T to a theory T' if in an obvious sense T' would be empirically worse than T?

In the empiricist tradition, Kemeny and Oppenheim noted this in a work of 1956, but in an incomprehensible way again sabotaged this observation yet in the same work. From the first sentence of their work onward, they connect the idea of reduction in an essential way with that of progress: "The label 'reduction' has been applied to a certain type of progress in science" (Kemeny and Oppenheim 1956, 6). And they tackle this connection from the very empirical side of things when they say: "As we see it, the essence of reduction cannot be understood by comparing only the two theories; we must bring in the observations" (ibid. 13). This occurs essentially in the form of the condition that, relative to a corpus O of well confirmed observational data, every part of O explained by T be also explained by T' is made a requirement for T to be reduced to T'. Here, the Hempel–Oppenheim notion of explanation is used for the, of course, singular statements of O. This means that the authors accept a hypothetico-deductive notion of corroboration for the empirical comparison of the two theories, and thus as part of their concept of reduction it is made clear that relative to O, the theory T' is empirically at least as well established as T, in the sense that all empirical successes of T in O are also those of T'.

So far, so good. But then the authors bother themselves at once with the dependence of their concept of reduction on the data set O and take it to disappear through a new and final notion of reduction in which they require their previous condition to apply for every O compatible with T: "for every conceivable [sic!] set of observational data ...consistent with T" (ibid. 18, Note 11). With this extension of a set of established observation statements to every conceivable such set, the newly recovered ground is immediately surrendered again. So, now, we are bound by the condition that all *possible* empirical successes of T as well as those of T'—and that is with respect to an actually present amount of data—that T' inherits all the actual success as well as all actual *failures* (as potential successes!) of T. Furthermore, Kemeny (1956, 17) seem to be aware of this when at the end they admit: "We are ignoring the fact that the theory to be reduced may be only approximately true ...We certainly would not want to require that an incorrect prediction of T should be a consequence of T'".

The implied notion of a merely *approximate reduction* was, as we have seen, given only somewhat half-heartedly by Hempel and Nagel, and Kemeny and Oppenheim saw difficulties here, from which they shrank (ibid. 13). The most committed among the philosophers in this matter has been *Popper*, whose falsificationist stance forced him, almost compulsively, to develop a conception of progress according to which newer theories throw their predecessors 'to the winds'. Exactly this, as Popper noted, is possible with a notion of approximate reduction (Popper 1949, 1958, 1975, 1979). As a hint to Bohr's correspondence principle shows (Popper 1958, 34), he also knew that, here, he had moved into the territory familiar to the physicists as well. Yet, he made no further use of it. His formulations remain similarly vague in wording as that of the physicists, and it remains in limbo, in particular, what the role played by observation statements in a concept of reduction is.

One thing in Popper is, nonetheless, as clear as can be. He conceives of a concept of theory explanation that will at the same time express the *progress* which occurs during the transition from an older to a newer theory that explains it. There is also a striking similarity between the way in which Popper argues his case and the view of Boltzmann and his followers with which we became familiar in Sect. 1.1. Popper distinguishes a progressive or *revolutionary* condition from a *conservative* one and sees the two together only as a feature of progress. In a later formulation (Popper 1975, 81–82), it is first said

> ...in order that a new theory should constitute ...a step forward it should conflict with its predecessor; that is to say, it should lead to at least some conflicting results. But this means, from a logical point of view, that it should contradict its predecessor: it should overthrow it. In this sense, progress in science—or at least striking progress—is always revolutionary.

But then there follows a condition requiring a certain continuity without which one can not speak of progress:

> My second point is that progress in science, although revolutionary rather than merely cumulative, is in a certain sense always conservative: a new theory, however revolutionary, must always be able to explain the success of its predecessor. In all those cases in which its predecessor was successful, it must yield results at least as good as those of its predecessor and, if possible, better results. Thus ...the predecessor theory must appear as a good approximation to the new theory.

It is evident that it remains open to what extent the reproduction of the respective older theory by its successor may (or must) take place: only a minimum condition with regard to the empirical data is set. The same holds for a formulation that describes the progress situation directly as an explanatory situation. Popper's prime example is the combination of Kepler's laws of planets with Galileo's law of falling bodies in Newton's theory of gravity. Concerning this, he says (Popper 1958, 33):

> Newton's theory unifies Galileo's and Kepler's. But far from being a mere conjunction of these two theories—which play the part of *explicanda* for Newton's—*it corrects them while explaining them*. The original explanatory task was the deduction of earlier results. Yet this task is discharged, not by deducing these earlier results but by deducing something better in their place: New results which, under the special condition of the older results, come numerically very close to these older results, and at the same time correct them.

It is obvious that, here, the strictly deductive concept of explanation—that of the 'the original explanation task'—is still present which he, of course, cannot handle, since the case at hand is one of theories contradicting each other (see Scheibe

1976a). Hence, only an approximate explanation is possible, to whatever extent it may however reproduce the older theory.

Popper has explicitly taken the view that (approximate) explanations can be deepened by iteration and that a corresponding progression of science occurs. But he has turned decisively against the possibility of *final* explanations (Popper 1958, 1974). Thus, without putting it this way, he was against the idea of the *unity of science*. Concerning these questions going far beyond the question of a possible unity even of physics, we have also to note a significant reluctance on the part of physicists. And we must attribute strong reservations even to Nagel, despite his proximity to empiricism and his reductionist attitude (Nagel 1949, beginning). It should be noted, however, that for logical empiricism in the strict sense the unity of science was almost programmatic—and was even regarded as the ultimate goal (Neurath et al. 1929). The idea of unity involves first of all the idea of *unification* of previously unconnected parts of science. That behind unification, especially of theories, there could even lie a final unity was certainly believed, for example, by Carnap. "In the history of physics"—he said (Carnap 1966, 242)—"it is always a big step forward when one branch of physics can be explained by another." After enumerating several examples, Carnap continued: "Slowly the notion grew that the whole of physics might one day be unified by one great theory." Of course, Camap still admits in this later text what he had to admit previously (Carnap 1938, 61): "there is at present *no unity of laws*. The construction of one homogeneous system of laws for the whole of science is an aim for the future development of science". And, when Camap then goes on to say that a unity of *the language* of science is already achieved, he can only claim so in the sense of a unified observation language—not in the sense of a unified theoretical language.

At the end of the classical phase of logical empiricism stands a work written in this spirit by *Oppenheim and Putnam* which boldly seeks to demonstrate through orderly reduction that contemporary empirical science presents an ontological hierarchy of empirical objects ranging from elementary particles to human society. The authors summarize the result of their investigation extending far into the natural sciences in the following words (Oppenheim and Putnam 1958, 28):

> It has been our aim ...to reply to the frequently made accusations that belief in the attainability of unitary science is 'a mere act of faith'. We hope to have shown that, on the contrary, a tentative acceptance of this belief ...as a working hypothesis is *justified*, and that the hypothesis is *credible*, partly on methodological grounds ...and partly because there is really a large mass of direct and indirect evidence in its favor.

Conceptually, Oppenheim (1958, 4) make intuitive use of an idea of what could be primarily meant by the unity of science:

> Unity of science in the strongest sense is realized if the laws of science are not only reduced to laws of some one discipline but the laws of that discipline are in some intuitive sense 'unified' or 'connected'. It is difficult to see how this last requirement can be made precise; and it will not be imposed here. Nevertheless, trivial realizations of 'Unity of Science' will be excluded. For example, the simple conjunction of several branches of science does not *reduce* the particular branches in the sense we shall specify.

This, in itself, old idea of a theory of inner unity, a unity into which in the process of unification the components merge, so to speak, has not been precisely explicated within logical empiricism, and even today we would be entering new territory with such an attempt.

So far, the resulting picture of the legacy of the philosopher in our subject, corresponds quite closely to that drawn by the physicists: Skepticism toward purely cumulative progress, but also skepticism—in addition to optimism—toward an eventual unity of physics, and finally a general endorsement of self-corrective, local progress which could perhaps be expressed by an appropriate concept of approximate reduction. This comparison would not turn out essentially differently if we were to also take into account the work of Lakatos (1978), who further developed Popper's falsificationism to a conception of progress which is realized in research programs.

The picture of the philosophers' legacy remains incomplete as long it leaves out the two major critics of the view sketched so far: Kuhn and Feyerabend. And it is only they with whom the two pictures approximately coincide—at least if one takes Heisenberg's little-noticed view just as seriously as the views of Kuhn and Feyerabend, which were honored with so much attention. Their deviation from orthodoxy begins with the rejection of the neo-positivist idea of a rational reconstruction of scientific events and leads to the idea of the *incommensurability* of scientific concepts and theories, as it characterizes the situation before and after the collapse of a scientific tradition and makes a relation of reduction impossible. Here, Kuhn has confronted the common view more to the historical facts, and Feyerabend more (and also) to what is desirable in terms of scientific progress. And, accordingly, the former believed in scientific revolutions only in the sense of the eventual replacement of a well-established tradition by another, while the latter instead insisted on a permanent plurality of theories.

Of the two, the historian *Kuhn* was thus the more conservative. His picture of the development of a scientific discipline was already suggested in the previous section in the words of Boltzmann, who anticipated it. The essential feature is the two-stage nature of normal science and of scientific revolution with a crisis serving as the transition. And revolutions are born from a conservatism of scientists acting together with the equal inexorability of its object (Kuhn 1977, 227):

> ...revolutionary shifts of a scientific tradition are relatively rare, and extended periods of convergent research are the necessary preliminary to them. As I will indicate ..., only investigations firmly rooted in the contemporary scientific tradition are likely to break that tradition and give rise to a new one.

Here, then, lies the reason for the two-stage nature of the model: Whoever wishes to talk about scientific revolutions must take into account the preceding periods of normal science. But, when a scientific revolution occurs, exactly what we know from Heisenberg happens: Not only are the laws of a theory gripped by it but, and primarily, its *concepts*. Kuhn commented on the usual argument for recovering Newton's mechanics from Einstein's as follows (Kuhn 1970a, 102, emphasis mine):

> It has ...not shown Newton's Laws to be a limiting case of Einstein's. For in the passage to the limit it is not only the forms of the laws that have changed. Simultaneously we had

> to alter the fundamental structural elements of which the universe to which they apply is composed. This need to change the meaning of established and familiar concepts is central to the revolutionary impact of Einstein's theory ... We may even come to see it as a prototype for revolutionary reorientations in the sciences. Just because it did not involve the introduction of *additional* objects or concepts the transition from Newtonian to Einsteinian mechanics illustrates with particular clarity the scientific revolution as a *displacement* of the conceptual network through which scientists view the world.

It is the situation thus described that the concept of incommensurability is intended to cover in general. Kuhn later termed a property of concepts and sentences such that it is the change of concepts and sentences with *this* property that has a revolutionary character, such that their counterparts in the respective new theory make this theory incommensurable to its predecessor (Kuhn 1977, 305; and Kuhn 1976, 198; emphasis mine):

> I suspect that, quite general, scientific revolutions can be distinguished from normal scientific developments in that the former require, as the latter do not, the modification of generalizations which had previously been regarded as *quasi-analytic*.

But, on the other hand, it must be stressed that a scientific revolution in Kuhn's sense need in no way be as radical as it is occasionally depicted. Although losses occur, there remains much ("though new paradigms seldom or never possess the capabilities of their predecessors, they usually preserve a great deal of the most concrete parts of past achievement," Kuhn 1970a, 169), new things are gained ("the new [paradigm] must seem to resolve some outstanding and generally recognized problem that can be met with in no other way," ibid.), and it only has to be borne in mind that the changes carry *more or less* the stigma of incommensurability. Even though Kuhn first coined this term frighteningly ("after a revolution scientists are responding to a different world," ibid., 111), he later presented it in a milder light (Kuhn 1983, 670–671), defining now properly:

> The claim that two theories are incommensurable is ...the claim that there is no language, neutral or otherwise, into which both theories, conceived as sets of sentences, can be translated without residue or loss.

But the damage balance usually only results in local difficulties:

> Most of the terms common to the two theories function the same way in both; their meanings, whatever those may be, are preserved ...Only for a small subgroup of ...terms ...do problems of translatability arise. The claim that two theories are incommensurable is more modest than many of its critics have supposed.

Systematically, Kuhn's notion of incommensurability of concepts and theories found an important addition in the scientific image of *Feyerabend*. A point of departure of this scientific image is indeed the conclusion that, what ever we may mean by progress, the admittedly *major* advances in science have always been made when theories incommensurable to existing beliefs have been established and enforced—such as the transition from the Aristotelian world view to Galilean-Newtonian physics, and again the turn from classical physics to relativity and quantum theory. Kuhn (1970a, 166) was amply pragmatic regarding the question of what progress through a revolution amounts to:

> Why should progress be the apparently universal concomitant of scientific revolution? Once again, there is much to be learned by asking what else the result of a revolution could be. Revolutions close with a total victory for one of the two opposing camps. Will that group ever say that the result of its victory will be something less than progress? ...and they are in an excellent position to make certain that future members of their community will see past history in the same way.

This is, in fact, the attitude of many scientists. Planck's words that while "a good theory will also be accepted in the absence of clever propaganda," it is not necessary thereby "that their opponents are convinced ...but rather that the opponents gradually die out and that the growing generation from the outset is familiar with the truth" are well known (Planck 1949, 10 and 13). A noteworthy formulation of this process is ascribed by Study (1914, VII) to the geologist Neumayr:

> It is a curious, constantly recurring phenomenon in the history of science: A new and correct view, which is not supported on new palpable material facts, but on a better interpretation of already known observations, is not attaining general acceptance because the opponents are convinced through the power of the reason ...but rather because they die out, and the young generation accepts the new theory as self-evident ...

Feyerabend (1975, 21) now considers, in opposition to the idea of the unity of science of the logical empiricists, a radical, i.e., not only methodological, *theory pluralism* according to which progress aims at horizon-broadening through introducing incommensurabilities:

> Knowledge ...is not a series of self-consistent theories that converges towards an ideal view; it is not a gradual approach to the truth. It is rather an ever increasing *ocean of mutually incompatible alternatives*, each single theory, each fairy-tale, each myth that is part of the collection forcing the others into greater articulation and all of them contributing, via this process of competition, to the development of our consciousness. Nothing is ever settled, no view can ever be omitted from a comprehensive account.

On the way toward this radical position, which he later adopted, Feyerabend had serious doubts regarding the applicability of the concepts of explanation and reduction of Hempel/Oppenheim and Nagel, respectively, to the theoretical development of physics. We have already seen that the condition of derivability presents difficulties, since it precludes that the reducing theory T' contradict the reducing theory T and that it thereby correct it. As to the solution of the merely approximate explanation or reduction, Feyerabend (1981a, 59) noted first that "dropping [the condition of (exact) derivability] means *giving up altogether* the orthodox theory, for [this condition] formed the very core of this theory" (emphasis mine). As much as one has to agree, Feyerabend's argument against approximate reduction (see also Feyerabend 1981b, 136–137), on the other hand, is unjustified, at least for the cases in which the other condition is not violated, which Feyerabend himself now targets directly: the condition of *meaning invariance.* In a reduction, the meaning of the concepts and sentences of the reduced theory are kept unchanged in the reducing theory:

> It is ...of utmost importance to note that expressions belonging to a science possess meanings that are fixed by its *own* procedures of explication ...whether or not the science has been reduced to some other discipline (Nagel 1961, 352).

The question is then how, under these circumstances, one is to understand, for example, the usual 'explanation' of the phenomenological gas equation via the kinetic theory, in particular, the 'explanation' of the phenomenological temperature via the average kinetic energy of the molecules. Feyerabend assumes that according to Nagel's 'natural' conception of reduction, none is actually available because the meaning of the word 'temperature' is not preserved. By contrast, he makes the claim "such a change to new meanings ...is a natural occurrence which is also desirable for methodological reasons" (Feyerabend 1981a, 81). But, in defense of Nagel, it must be said that this is made clear enough, despite some already mentioned uncertainty in the matter, when he expressly requires as part of heterogeneous reduction not the above quoted passage invoked by Feyerabend, but rather his 'conditions of connectability'. And Nagel (1961, 357) has not overlooked the fact that heterogeneous reduction *should* provide a change of meaning.

> It is certainly possible to redefine the word 'temperature' so that it becomes synonymous with 'mean kinetic energy of molecules'. But it is equally certain that on this redefined usage the word has ...a meaning different from the one associated with the word in the statement of the Boyle-Charles' law. However, if thermodynamics is to be reduced to mechanics, it is temperature in the sense of the term in the classical science of heat which must be asserted to be proportional to the mean kinetic energy of gas molecules.

But, with the problem of meaning invariance, it is not even primarily about the fate of old concepts in heterogeneous reductions that can, incidentally, be performed without difficulty. The main problem is, again, that of the incommensurability of concepts and theories (that contain them). Now Feyerabend (1978, 179–180), differently from Kuhn, gives this a truly formidable appearance:

> *Mere difference* of concepts or of theories ...does not suffice. It is also not enough to show that certain terms of the first theory can not be combined with certain terms of the second theory. In fact, the situation must be such that the use of just *one* sentence or just *one* term of a theory makes use of *all* the sentences and all other terms of theory impossible.

Feyerabend's notion of incommensurability (like Kuhn's) is a notion of incompatibility. But it is not the typical logical case of inconsistency familiar to us, the way Popper rendered it as the core of theoretical progress. Two conflicting theories can (and even must), nevertheless, have a common interpretation; it is just that no such interpretation can make both theories *true*. Two incommensurable theories with the same reference, however, even fail to have a common *interpretation*. Even the interpretation of one of the two supplies, in a certain sense, so complete an understanding of the matter that nothing more can be added. In this sense, interpretive inconsistency is more thoroughgoing than the other, logical incompatibility.

Concerning possible instantiations, one might thus at first consider something like the wave-particle duality of the (older) quantum theory, and so consider a relationship of incommensurability with Bohr's concept of complementarity (see Scheibe 1988b). For, indeed, wave and particle pictures don't imply contradictions, but conflict through the disparate meaning of their basic concepts (which is certainly what Feyerabend (1981b, 446), means even if he says the opposite). Feyerabend is also the only philosopher of science, it seems, who cites Bohr's warnings, with which we are familiar from the previous section and which arise from the peculiarity of his complementarity concept (ibid., 372–382). That would also be compatible with the somewhat different view that incommensurability is about the relationship between theories, one of which, the older, has concepts which have suffered a shift in meaning in the newer, more recent one (Feyerabend 1978, 178–182). Feyerabend exemplified such relationships by the cases of the transitions from pre-relativistic to relativistic physics (Feyerabend 1965), of Newton's to Einstein's theory of gravitation (Feyerabend 1981b, 140–142), and from classical to quantum mechanics (Feyerabend 1981b, 445–453, and Feyerabend 1978, 186–191). In contrast to the above example of Nagel, where the concepts of the old theory are missing in the new theory, in these cases the relevant successors to the old concepts are clearly present, but are connected in such intricate ways within the new theory that the old theory can be obtained from the new only with great difficulty, if at all. Thus, in the above mentioned cases, the concepts—e.g., the length of a distance and the simultaneity of two events in the first case, the metric and the gravitational field in the second, and the position and momentum in the third—have unique successors (of the same names!) in the successor theory. But these successor concepts are connected there in a completely different way from their predecessors in their theories and, thereby, greatly complicate the recognition of the old in the new.

We are faced with similar problems as we had when trying finally to complete Heisenberg's analysis of a revolutionary theory change using a notion of progress. The relationship to Heisenberg's conception of the closed theory arises from the fact that, according to Feyerabend, incommensurability is impossible without both of the connected theories being *universal* in a fairly thoroughgoing sense, i.e., without it being the case that they "contain the means for the description of every process possible in their respective domain" (Feyerabend 1978, 306–307). Universal theories are, therefore, already incommensurable if they are different. And this idea is also suggestive in the case of closed theories. It is only that for Heisenberg, by contrast, here the characteristic *property* is the key concept, whereas for Feyerabend it is the *relationship* of theories. (Kuhn's notion of quasi-analytic propositions of a theory also belongs here.) In any case, the concepts of closedness and incommensurability have become the greatest known challenge for a conception of physics, which is marked by the concepts of continuity, progress, reduction, unity, etc., as means of description. Those difficult concepts may have been insufficiently thought through by their protagonists and exaggerated in significance. But, from the opposite side of orthodox philosophy of science, they received too little attention or were simply undervalued. It is time for a balanced view of the matter.

1.3 The Program

We must immediately make a restriction to this aim, if it is a restriction,—namely, that we will not agree or will agree little with the critics of the orthodox philosophy of science on another point: their criticism of the idea of *rational reconstruction*.

Philosophers such as Carnap, Reichenbach, Hempel and other logical empiricists were basically in agreement among themselves and even with their internal critic Popper (see Scheibe 1984a) that there is, besides the history, psychology, and sociology, etc., of science, something like its rational reconstruction by logico-philosophical analysis, and that it is to be considered the legitimate subject of independent philosophy of science. Thus, we find in Carnap the characterization of his constitutional system in 1928 as "a rational reconstruction of the entire formation of reality, which, in cognition, is carried out far the most part intuitively" (Carnap 1928, 158). Already here, the manifest psychological and, from a rational point of view, disqualifying use of the word 'intuitive' indicates what Carnap expresses in a different formulation more precisely, "that the construction does not represent the actual process of cognition in its concrete manifestations, but that it is intended to give a rational reconstruction of the formal structure of this process" (ibid., 220). A few years later, Popper took up Carnap's neologism for the realm of what is generally rational reconstructable, expressly limited to the final stage of a process of knowledge, namely, the *examination* of the idea of a researcher: "In so far as the scientist critically judges, alters, or rejects his own inspiration we may, if we like, regard the methodological analysis undertaken here as a kind of 'rational reconstruction' of the corresponding thought-processes. But this reconstruction would not describe these processes as they actually happen: it can give only a logical skeleton of the procedure of testing" (Popper 1935, 8). Fundamentally along the same lines, a few years later, Reichenbach (1938, 5–6) in turn said:

> Epistemology does not regard the processes of thinking in their actual occurrence; this task is entirely left to psychology ...Epistemology ...considers a logical substitute rather than real processes. For this logical substitute, the term *rational reconstruction* has been introduced ...It is ..., in a certain sense, a better way of thinking than actual thinking. In being set before the rational reconstruction, we have the feeling that only now do we understand what we think.

And, still decades later, Carnap again explained anew his old concept, to the effect that a rational reconstruction is "the searching out of new definitions for old concepts. The old concepts did not ordinarily originate by way of deliberate formulation, but in more or less unreflected and spontaneous development. The new definitions should be superior to the old in clarity and exactness...Such a clarification of concepts...still seems to me one of the most important tasks of philosophy, ..." (Carnap 1969, v).

One can interpret the reconstructionism indicated by these quotes narrowly or broadly. Carnap was an extreme case of the former direction. His narrowness was primarily in his formalist attitude. A broader reconstructivistic position was the demarcationism of Lakatos (1978, Vol. 1, Chap. 2, and vol. 2, Sect. 6) who followed Popper. But Lakatos was already in an explicit confrontation with the critics of this entire direction. The critics attacked mainly the history of knowledge aquisition and, in particular, the history of science, and tried to show that decisive steps in the development of physics are not found in the rational reconstructions of the philosophers of science, while, conversely, these reconstructions are not to be found in the actual reality of science. Also on this level, it is above all again Kuhn (1977, 14) who expresses his discontent with the usual purely philosophical analysis. Mainly based on textbook presentations and, historically, mainly from a few classics of science—Galileo, Newton, Darwin, etc.—"the philosopher's reconstruction is generally unrecognizable as science to either historians of science or to scientists themselves." Here, Kuhn's critique is not directed against rational reconstructions as such: also, both "historians and scientists can claim …to be as concerned with essentials, to be engaged in rational reconstruction. Instead the difficulty is the identification of essentials. To the philosophically minded historian, the philosopher of science often seems to have mistaken a few selected elements for the whole and then forced them to serve functions for which they may be unsuited in principle and which they surely do not perform in practice, …" (ibid.). Thus, Kuhn holds historically relevant rational reconstructions of science to be quite possible. Nevertheless, he demands of his own reconstructions just what orthodox reconstructionism expressly seeks to abolish:

> The explanation [of scientific progress] must, in the final analysis, be psychological or sociological. It must, that is, be a description of a value system, an ideology …Knowing what scientists value, we may hope to understand what problems they will undertake …I doubt that there is another sort of answer to be found (Kuhn 1970b, 21).

In contrast to this apparently rather irreconcilable position, a logical, in the broad sense, reconstruction of certain aspects of physics, including those that concern its development, will be considered in this work to be quite possible and has its own merit, not as a competitor, but as a *complement* to a purely historical one and, as such, psychological or sociological studies. 1960s and 1970s put forth a historical claim to absoluteness on the part of critics of logical reconstructionism that cast doubt on the legitimacy of all non-historical questions. Certainly, a historian or philosopher of science acting as such can be interested just in the development of scientific concepts in a stage in which the vagueness of these concepts defies any attempt at their logical reconstruction. Conversely, current demands on an adequate formulation of mechanics may contribute nothing more to our understanding of Newton's 'Principia'. Even though a logician wouldn't dream of blaming a historian for having neglected logic, it has recently become popular to blame logical reconstructionists for their historical ignorance. But, as important it is to understand history and as gratifying it is to see compensation for the neglect of the historical dimension in the 1930s until the 1950s, everything that has occurred in the meantime has provided no reasons for subjecting the philosophy of science to a pan-historistical monopoly.

I will now not mention that the opponents of logical-systematic reconstructions frequently forget their historical message when logic comes into play as if logic did *not* belong to the historical reality of our science (e.g., Feyerabend 1975, 190–197). Aside from this historical integration, in the respective scientific areas there is, above all, a systematic continuity that does not allow certain parts to be separated from others as irrelevant without considerable arbitrariness. One can draw lines, starting from the concreteness of our sensations up to the abstractness of logical inference, that can be truncated at any point by the same right or wrong as any other. Let's take, for example, our sensations of heat which are explicated in a first step by measuring instruments for temperature. The logical empiricists have said that here the colloquial terms of hot and cold are explicated via the concept of temperature.

Naturally, this has absolutely nothing to do with philosophy. Rather, we are now only in experimental physics, and this reconstruction is its account. But, then, we have theoretical physics, which tells us in the kinetic theory of gases (in its simplest form) that the temperature is the average kinetic energy of the molecules. Thus, the concept of temperature is introduced into a theoretical network and opens up the possibility of determining even remote temperatures: through measurement *and* calculation. At this stage of the reconstruction of the concept of temperature, much mathematics is already in play and—in the final step—logic. In particular, there is no indirect determination of the temperature or any physical quantity without finally calculating and, thereby, deducing the value.

Where should one now non-arbitrarily put a full-stop here before having at least attained logic? Mankind got by long enough without thermometers, and still today there are experimental physicists who face theories with extreme skepticism, again and again one encounters theoretical physicists who find themselves willing only to the most sparing use of mathematics, and many mathematicians have wanted and want to know nothing of logic. From this, we see that, in the current philosophical attempts to lend a precise logical status to certain metaphysical terms, nothing happens here that is simply outrageous, that would entitle us to enforce a legitimation obligation which would not be appropriate anywhere else with the same right. Philosophy of science is not a "hitherto unknown form of madness" (Feyerabend 1978, 293), but has its legitimate place in the midst of the vast landscape of our sciences as a whole. So let us get to work and affirm its place!

This is not intended here now to be accomplished by a general analysis and defense of rational reconstructionism. Because the only thing that matters here is to confirm the idea of a logical reconstruction actually by its successful application—in our case, in the realm of reduction, progress, and unity of physics. If the method fails for a subject like this, then any general talk serves no purpose, and if one has success with it, then such talk adds nothing to it. Put up or shut up—that's the motto. As will immediately become clear, we will, however, in the analysis of our main subject—the reduction of theories—deviate from the usual methods of philosophy of science already on the general level and, to that extent, indirectly criticize these methods. Also, the following treatment of the subject, contrary to the common practice of philosophy of science, is guided to an extraordinary extent by concrete examples, which are however not historical in a strict sense. But this does not change the principal recourse to the classical idea of rational reconstruction through logical analysis.

The main theme of this work is the *reduction* of a physical theory to another—reduction understood in the literal sense of tracing back (cf. the introduction of Scheibe 1993c and Scheibe 1995a). In answering the question of what a reduction in physics is, we will not—the deviation begins already here—take the usual route of an explication, which leads from an explicandum to adequacy conditions and finally to a definition (see Carnap 1950, Chap. I, and Scheibe 1984a for the older treatment already in Kant). Nevertheless, we accept as a kind of guiding principle a *preliminary notion* of reduction, to which we shall turn our attention first. The question is, whether we wanted to set up a theory inventory for physics, specify without exposing ourselves to the criticism that *either* we have forgotten a theory, so that our list would be incomplete, *or* we have invoked a theory too much, so that our list would be uneconomical. If we now consider how to prevent the first error (as such) or how the second could be counted (as an error), it is in both cases the same argument: We would argue that the missing or superfluous theory can be *traced back* to the other theories—it is somehow already present in them, and thus there is redundancy in the list. The question of what a theory reduction is, therefore, boils down to the question of which methods of obtaining theories from other theories we consider in physics as such methods, ones by which the respective 'deduced' theory, in principle, is not a new one that is independent of the initial theories.

Thus we encounter here a situation that is already well known in the explanation and prediction of individual events: Just as any such explanation, apart from its pragmatic circumstances, has the same logical structure as a prediction and could have been a prediction, so every reduction of an older theory to a newer (and better) has the logical structure of a 'deduction' and, in this respect, also could have recovered a previously unknown theory from a once known (and confirmed) theory. With the previously discovered laws of Kepler, Newton was in a position where he needed to trace these laws back to his own theory of gravity. After already having his theory, it was natural to want, for example, to treat the sun-earth-moon system, which Kepler had not considered, with a Newtonian theory as simple as possible that neglects the interaction between the sun and moon. One, then, legitimately attributes the success of this treatment to the original Newtonian theory, if only the actual conditions permit the approximation. Such 'deductions' of 'new' theories, which employ approximations, today pervade almost all parts of physics. As with explanation and prediction, the situation is not symmetrical: Just as not all predictions cannot (subsequently) be treated as explanations (see Hempel 1965, 367–371), so not all methods of derivation have the strength of a reduction. And this is true even for the derivation in the strict sense, i.e., for purely logical deductions from a theory (without any addition or diminuition). Already on the basis of the first-mentioned, other direction, however, the scope of application of the relevant derivation methods does include, in addition to actual reductions of a theory T to a theory T', also the 'derivations' of T from T' (in a preliminary sense of 'derivation'). Therefore, clarifying the question of which methods are employed here is not only relevant for the backward-turned reformulation of physics but for its forward-looking practice of research.

Having a preliminary conception of reduction, we can now attempt to build up a theory. It is from here on that the following proposal differs from the usual procedure in such cases, that is, the analytical approach. Usually, we would replace the still vague preconception by a more precise explication with the aid of so-called adequacy conditions, i.e., the properties which we want to achieve in any case for our concept. By contrast, the theory to be provided here does almost the reverse. While the adequacy conditions would, according to their determination, have to be *more* general than the concept to be explicated, we do not follow an analytical but rather a *synthetic* approach and, thereby, gain a general notion of reduction, by starting from as specific as diverse reductions and, through iterated *combination*, put forward new and ever more complicated reductions. Fundamental for the viability of this approach is the fact that we can compose (or combine) reductions: If we have reduced theory T to theory T', and T' in turn to the theory T'', then we have reduced T to T''. This, in itself, trivial fact not only serves to put in operation the subsequent theory of reduction, but also makes the deepening of reductions possible.

The possibility of successive reductions (or, 'derivations') leads to the idea of viewing this concatenation as a *multiplication*, such as the transformations or even the maps of a set into itself, combined to a group or, respectively, a semigroup. This view is certainly of at least heuristic value, but one must ensure a priori that the multiplication of any two *arbitrary* elements, such as the algebra usually provides, appears here not for the reductions themselves but only for the *kinds of reduction* in question. The distinction between the individual reduction, on the one hand, and the reduction type to which it belongs, on the other, is actually of importance for the understanding of the matter. In the kinds of reduction, we abstract from the theories involved in reduction, while the reductions themselves naturally contain both theories taken as reducing partners. Thus we speak, on the one hand, for example, of the reduction of Kepler's laws to Newton's theory of gravitation and virtually name the reduction after the theories that are connected by the reduction. But, when it is asked what kind of reduction is thereby accomplished, then we are asked for the reduction type that lies before us, and we respond with denominations such as 'generalization', 'limiting case reduction', or whatever else it is, without mentioning the theories. The consideration of two successive reductions now makes it immediately clear that to combine them arbitrarily is out of the question. We must at least satisfy the connecting condition that the theory obtained by the first reduction is *the same* theory that is reduced in the second reduction. In contrast, it appears that there is no obstacle to the possibility of combining different *kinds* of reduction, and so we can speak of a multiplication of these elements—a multiplication that is moreover associative. Of course, the kind of the product of two reductions (if it can be made) is identical to the product of the associated kinds of reduction.

After this clarification, it should be emphasized once again that the combining of reductions and reduction types is not a casual accessory of our approach—it will be the decisive factor which allows the enterprise of reduction to flourish. In fact, it can be demonstrated in physics to a half dozen iterations without difficulty. But, because we must start with something, the question of *elementary reductions* arises. Elementary reductions—or better, reduction types—could, for example, only trivially be combined from others. We know this from the natural numbers, for which every natural number is obtained by multiplying prime numbers, even though the prime numbers are (multiplicatively) irreducible and establish this recursion. However, with reductions, the trivial decomposition is not characterized by an identity element, but by idempotence: The only manner of representing an elementary reduction type A as a combination is A^2, i.e., if $A = B \cdot C$, then $B = C = A$, and so $A = A^2$. The elementary reductions are mentioned here mainly for heuristic reasons: It is obvious that the possible definition just mentioned makes sense only if one has already *independently* established the range of possible reductions, as in the number analogy and the concept of natural number via the successor relation. Since such a definition is unknown in our cases, we will hereafter simply start with certain reduction types, for which no genuine decomposition is known, and form combinations of them. The universe of possible reduction types is therefore *open* in the logical sense, something fortunate for a business that is as yet temporally incomplete. The specific drawback that may arise from this openness is far outweighed by our synthetic method, which—in contrast to general conceptions—is not in need of correction if new elementary reduction types are found and permitted.

Before we come finally to notions related to the concept of reduction, we have to add another novelty to the theory. The usual treatments of our subject and, in fact, the usual way of speaking suggests that reduction is about a two-place predicate: a theory T is reduced to a theory T', and that's all there is to it. In reality, however, there is a third element of such importance that one could almost say that it is precisely this third element that accomplishes the reduction. In the reduction of Kepler's laws to Newton's law of gravitation already mentioned, the third element is the contingent conditions under which Newton's law alone gives rise to Kepler's laws, i.e., that the mass of the sun is great, that the distances of the planets are not too small, etc. It is obvious that, here, without such related conditions, there would be no reduction at all. And that is not so only here, but there is for each reduction type a characteristic *vehicle*, as it should be called, which first brings about the reduction.

So much for the concept of reduction for the moment. The word 'reduction' as it appears in Sect. 1.1 does not appear at all in the work of physicists themselves. It is paraphrased in various ways, sometimes one speaks of an explanation, and it is clear that there the subject should somehow be captured through a *notion of progress*. Progress in relation to the natural sciences is now well understood primarily as technical progress. Any improvement would belong to it, and, in particular, any refinement of measurement methods, as it is always been necessary for the establishment of theoretical progress. Indeed, it may exactly be that the interaction between theoretical and experimental physics is crucial for the development of physics as a whole. Here, not only does refining the measurement technology make a theory successful, but conversely new theories also indicate to experimental physics what and how to measure. Of this integral phenomenon, only *theoretical* progress will, in the following, be of concern to us: A physical theory T' may be an advance over another theory T, such as Newton's celestial mechanics over Kepler's laws, or electrodynamics to electrostatics. Each advance is, thus, one of theory, in so far as what are connected are theories. This emphasis on theory before experiment now brings qualities into play which shall not be treated here. Thus, the theory T' can have the advantage over T of being simpler, and so a transition from T to T' would certainly be a theoretical advance in this respect. But questions of the simplicity of theories and related questions which ultimately interfere more with aesthetics are difficult to handle and should remain completely excluded here. But we must be concerned with the quality of theoretical progress that shows as *empirical* progress that a theory T' does demonstrably better justice to its task to reproduce physical reality than does a theory T—that to a certain extent, T' gives rise to better empirical data than T.

Progress that is both empirical and theoretical in the indicated sense has nothing directly to do with reduction. The decision as to whether the theory T' implies progress compared to theory T should depend on nothing beyond the applications of T and T' up to a certain point in time. We could provisionally say that T' is an empirical advance over T if T' can explain every empirical phenomenon that T already explained and if it explains further phenomena that T could not explain and, particularly, had not properly explained. In other words, T' must reproduce the empirical success of T and, beyond that, have new empirical successes, especially those that were failures for T. That sounds quite plausible, but a proof that T' is an advance over T in this sense can pose considerable difficulties. The main difficulty may be to relate the phenomena relevant to T', on the one hand, and T, on the other hand, *all* to each other—*before* any success can be recorded. We recall here, once again, a quote from the first section in which Nernst (1922, 489) regards the relationship of Newton's to Einstein's theory of gravitation, wherein he says that "as a matter of principle every computation that astronomers have performed so far must be changed." Here "computation ...performed" means the calculations carried out in order to test Newton's theory. And the requirement to change them means that the same empirical phenomena (as the respective connection of empirical data) had to be explained in Einstein's theory by *different* calculations—confirming Newton's as well as refuting it. Indeed, the substantial differences between the theories could imply for a phenomenon formulated in Newton's theory that it has no clear counterpart in Einstein's theory, and vice versa, so that empirical comparison stalls already for this reason.

Considering this difficulty explicitly gives rise to the idea that there could be some theoretical relations that make it possible to separate the problem of empirical comparison of theories from the particular matters of fact, i.e., the actual empirical successes and failures because these theoretical relations *themselves* represent global guarantees, if only potential ones, of empirical progress. Reduction now offers itself as just such a theory relation: We may expect that a theory T' which can be traced back to a theory T is at least not empirically worse than T. Of course, regarding two theories T and T' that are related to each other by a reduction, just thereby we know nothing at all about their absolute empirical usefulness. The question can only ever be whether empirical success of T, *if and so far as* available, can be carried on to T' and whether potential failures can at least already be mitigated *due to* a reductive relationship between T and T'. Even this will, of course, only be the case if T and T' are theories that can separately be empirically tested. Thus, although we keep in view this possible link between reduction and empirical progress because of its fundamental importance, and almost regard the possibility of empirical progress from T to T' as a necessary condition for the possibility of reducing T to T', we must already point out here that this is a particularly difficult area which can not be significantly advanced by this work.

In possession of the notions of reduction and progress, we are also now capable of illuminating the concept of the *unity* of physics somewhat. A physical theory represents a unity with respect to the multiplicity of theories 'derived' from it or, equivalently, to which they can be reduced. Of course, this is only a relative concept of unity, moreover, that still requires nothing positive of a theory. In particular, it remains completely open how comprehensive a theory is in a more absolute sense. But our notion covers at least what we know from the history of physics of the processes that we would call the 'unification' of several theories into a single one. Thus, Newton's theory of gravitation unified Kepler's laws and Galileo's law of falling bodies, and electrostatics, magnetostatics and optics have been unified by Maxwell's equations. In such cases, nothing happens other than that the previously well-established theories can be reduced to *the same* new and independently confirmed theory. Of course, it initially remains completely open how the unification obtained looks within the wider landscape of physical theories. It is certain that unifications in the sense of multiple reductions become of greater value with the more theories they unify. And, of course, multiple (reduction-equivalent) 'derivations' from a *single* theory that has once been obtained are as valued as those multiple reductions.

In all these cases, the use of the word 'unity' is legitimate and would already describe a procedure all the way up to the unity of physics as a whole, without having given the concept of unity already a *unique character*. But, it should be assumed that, even with this simple process of reductive unifications, such a special quality plays a role—whether we know it or not. It is this difference between mere unification and unity in the sense of a characteristic quality that Oppenheim and Putnam (1958, 4) intend when they, as cited in Sect. 1.2, state that:

> Unity of science in the strongest sense is realized if the laws of science are not only reduced to laws of some one discipline but the laws of that discipline are in some intuitive sense 'unified' or 'connected'.

The big problem is that of which precise, not merely intuitive sense the expressions 'unified' or 'connected' should be understood here. Rationalist systems that have the comprehensibilty of the world as a main thesis have repeatedly attempted to make this thesis credible through world schemes in which everything is connected with everything and maximum conceptual coherence is achieved. But to enforce this sort of high claim, for example, a coherence theory of truth is then used, according to which truth comes in degrees, and such constructions have not been based on solid logical foundations compatible with the physics that we currently have. At the moment, they cannot be included in a rational reconstruction of physics (Scheibe 1994a). In this treatise, therefore, we will—like Oppenheim and Putnam—reach little beyond the concept of unification associated with multiple reductions, and leave out anything going beyond that, more or less, to intuition.

References

Balzer, W., Moulines, C. U., and J. D. Sneed. 1987. *An Architectonic for Science. The Structuralist Program*. Dordrecht: Reidel.

Bohm, D. 1980. *Wholeness and the Implicate Order*. London: Routledge & Kegan Paul.

Bohr, N. 1934. *Atomic Theory and the Description of Nature*. Cambridge, UK: Cambridge University Press.

Bondi, H. 1977. The Lure of Completeness. In: R. Duncan and M. Newton-Smith (Eds.), *The Encyclopedia of Ignorance*. Oxford: Pergamon Press, 6–8.

Boltzmann, L. 1905. *Populäre Schriften*. Leipzig: J. A. Barth.

Boltzmann, L. 1979. *Populäre Schriften* (1905). New Edition by E. Broda, Braunschweig: Vieweg 1979.

Brittan, Jr., G. G. 1970. Explanation and Reduction. *J. of Phil.* LXVII, 446–457.

Carnap, R. 1928. *Der Logische Aufbau der Welt*. Leipzig: Felix Meiner. Quoted after the Engl. transl. by R. A. George: *The Logical Structure of the World. Pseudoproblems in Philosophy*. (2nd Printing; 2005) California: University of California Press.

Camap, R. 1938. Logical Foundations of the Unity of Science. In: *Int. Encyclopedia of Unified Sciences*, Vol. I, 42–62.

Carnap, R. 1958. Beobachtungssprache und theoretische Sprache. *Dialectica* 12, 236–247. Engl. transl.: Observational Language and Theoretical Language. In: J. Hintikka (Ed.), *Rudolf Carnap, Logical Empiricist: Materials and Perspectives*. Dordrecht: Springer Netherlands 1975.

Carnap, R. 1950, ^{2}1962. *Logical Foundations of Probability*. Chicago: University of Chicago Press.

Carnap, R. 1966. *Philosophical Foundations of Physics*. New York: Basic Books.

Carnap. R. 1969. R. A. George Ttransl.), *The Logical Structure of the World. Pseudoproblems in Philosophy*. (2nd Printing; 2005) California: University of California Press.

Causey, R. L. 1977. *Unity of Science*. Dordrecht: Reidel.

Dilworth, C. 1981. *Scientific Progress*. Dordrecht: Reidel.

Dirac, P.A.M. 1929. Quantum Mechanics of Many-Electron Systems. *Proc. Roy. Soc. of London*, Ser. A, vol. CXXIII, 714–733.

Dirac, P.A.M. 1966. *Lectures on Quantum Field Theory*. New York: Belfer Graduate School of Science.

Einstein, A. 1914. Antrittsrede. *Sitzungsberichte Königl. Preuss. Akad. Wiss.* XXVIII (1914), 739–742.

Einstein, A. 1917. *Über die spezielle und die allgemeine Relativitätstheorie*. Vieweg: Braunschweig ^{23}1988.

Einstein, A. 1919. My Theory. *The Times*, 28 Nov 1919. Repr. in *The Living Age* (3 Jan 1920), 304, 41–43. https://todayinsci.com/E/Einstein_Albert/EinsteinAlbert-MyTheory.htm

Einstein, A. 1936. Physik und Realität. Journal of The Franklin Intitute 21, 313–347. Engl. transl.: Physics and Reality. Ibid., 349–382.

Einstein, A. 1950. On the Generalized Theory of Gravitation. *Sci. Amer.* 182, 13–17.

Feyerabend, P. 1965. On the "Meaning" of Scientific Terms. *J. of Phil.* LXII, 260–274.

Feyerabend, P. 1970. Problems of Empiricism, Part II. In: *The Nature and Function of Scientific Theories*. R. G. Colodny (Ed.) Pittsburgh: University of Pittsburgh Press, 275–353.

Feyerabend, P. 1975, ^{3}1993. *Against Method. Outline of an Anarchistic Theory of Knowledge*. London: Verso, 1975. Quoted after the 3rd edition which is based on the revised and extended German edition: *Wider den Methodenzwang. Skizze einer anarchistischen Erkenntnistheorie*. Frankfurt a. M.: Suhrkamp 1976 (here: 48–49 and 346–357).

Feyerabend, P. 1978. *Der wissenschaftstheoretische Realismus und die Autorität der Wissenschaften*. Braunschweig: Vieweg.

Feyerabend, P. 1981a. Realism, Rationalism and Scientific Method. Philosophical Papers Volume 1, Cambridge, UK: Cambridge University Press.

Feyerabend, P. 1981b. *Probleme des Empirismus: Schriften zur Theorie der Erklärung, der Quantentheorie und der Wissenschaftsgeschichte*. Ausgewählte Schriften Bd. 2. Braunschweig: Vieweg.

Friedmann, K. 1982. Is Intertheoretic Reduction Feasible? *Brit. J. for the Phil. of Sci.* 33, 17–40.

Guillemin, V., and S. Sternberg 1984. *Symplectic Techniques in Physics*. Cambridge, UK: Cambridge University Press.

Hawking, S. 1980. *Is the End in Sight for Theoretical Physics?* Cambridge, UK: Cambridge University Press.

Heisenberg, W. 1952. *Philosophic Problems of Nuclear Science. Eight lectures*. London: Faber and Faber. German original: *Wandlungen in den Grundlagen der Naturwissenschaft*. 2nd ed.: Leipzig: Hirzel 1936 (here: 25; 12th ed.): Stuttgart: Hirzel 2005 (here: 46).

Heisenberg, W. 1971. *Physics and Beyond: Encounters and Conversations*. New York: G. Allen & Unwin 1971. German original: *Der Teil und das Ganze. Gespräche im Umkreis der Atomphysik*. München: Piper 1969.
Heisenberg, W. 1974. *Across the Frontiers*. New York: Harper and Row 1974. German original: *Schritte über Grenzen*. München: Piper; here: 306–313.
Hempel, C. G. 1965. *Aspects of Scientific Explanation and Other Essays in the Philosophy of Science*. New York: The Free Press.
Hempel, C. G. 1966. *Philosophy of Natural Science*. Englewood Cliffs, NJ: Prentice-Hall.
Hempel, C. G., and P. Oppenheim 1948. Studies in the logic of explanation. *Philosophy of Science* 15, 135–175.
Hendry, J. 1984. *The Creation of Quantum Mechanics and the Bohr-Pauli Debate*. Dordrecht: Reidel.
Hertz, H. 1892, ^{3}1914. *Untersuchungen über die Ausbreitung der elektrischen Kraft*. Leipzig: J. A. Barth.
Hilbert, D. 1919. *Natur und mathematisches Erkennen. Lectures Göttingen 1919-1920.* From notes by P. Bemays. Ed. by D. E. Rowe. Springer: Basel 1992.
Kangro, H. 1970. *Vorgeschichte des Planckschen Strahlungsgesetzes*. Wiesbaden: Steiner.
Kemeny, J. G., and P. Oppenheim. 1956. On Reduction. *Phil. Studies* 7, 6–19.
Krajewski, W. 1977. *Correspondence Principle and Growth of Science*. Dordrecht: Reidel.
Kuhn, T. S. 1962, ^{2}1970. *The Structure of Scientific Revolutions*. Chicago: Univ. of Chicago Press.
Kuhn, T. S. 1970b. Logic of Discovery or Psychology of Research? In: Lakatos and Musgrave 1970, 1–23.
Kuhn, T. S. 1976. Theory Change as Structure Change: Comments on the Sneed Formalism. *Erkenntnis* 10, 179–199.
Kuhn, T. S. 1977. *The Essential Tension: Selected studies in scientific tradition and change*. Chicago and London: The University of Chicago Press. German Transl.: *Die Entstehung des Neuen*. Hrsg. L. Krüger. Frankfurt am Main: Suhrkamp 1977.
Kuhn, T. S. 1983. Commensurability, Comparability, Communicability. In: P. D. Asquith and T. Nickles (Eds.), *PSA* , vol. 2. East Lansing, Mich., 669–688.
Lakatos, I. 1978. *Philosophical Papers*. 2 vols. Ed. by J. Worrall and G. Currie. Cambridge, UK: Cambridge University Press.
Lakatos, I., and A. Musgrave (Eds.) 1970.*Criticism and the Growth of Knowledge*. Cambridge, UK: Cambridge University Press.
Ludwig, G. 1990. *Die Grundstrukturen einer physikalischen Theorie*. Berlin: Springer.
Mayr, E., and S. Weinberg. 1988. The Limits of Reduction. *Nature* 331, 475–476.
Misner, C. W., Thorne, K. S., and J. A. Wheeler 1973. *Gravitation*. San Francisco: Freeman.
Nagel, E. 1949. The Meaning of Reduction in the Natural Sciences. In: R. C. Stauffer (ed.), *Science and Civilization*. Madison, WI: University of Wisconsin Press, 99–135 (reprinted in: A. Danto and S. Morgenbesser (Eds.), *Readings in the Philosophy of Science*. New York: Meridian 1960, 288–312).
Nagel, E. 1961. *The Structure of Science*. New York: Harcourt, Brace & World.
Nagel, E. 1970. Issues in the Logic of Reductive Explanations. In: H. E. Kiefer and M. K. Munitz (Eds.), *Contemporary Philosophic Thought. Mind, Science, and History*. Albany, N.Y.: State University of New York Press, 117–137.
Nernst, W. 1911. *Theoretical Chemistry*. London: Macmillan 1911. Transl. of: *Theoretische Chemie*. Stuttgart 1893.
Nernst, W. 1922. Zum Gültigkeitsbereich der Naturgesetze. *Naturwissenschaften* 10, 489–495.
Neurath, O., Carnap, R., and H. Hahn 1929. *Wissenschaftliche Weltauffassung—Der Wiener Kreis*. Wien: Wolf.
Nickles, T. 1973. Two Concepts of Intertheoretic Reduction. *J. of Phil.* 70, 181–201.
Niiniluoto, I. 1984. *Is Science Progressive?* Dordrecht: Reidel.

Oppenheim, P., and H. Putnam. 1958. Unity of Science as a Working Hypothesis. In: H. Feigl et al. (Eds.), *Minnesota Studies in the Philosophy of Science*, vol II. Minneapolis: Univ. of Minnesota Press, 3–36.

Pauli, W. 1979. *Wissenschaftlicher Briefwechsel* I. Ed. by A. Hermann et al. Berlin: Springer.

Planck, M. ^{5}1949. *Vorträge und Erinnerungen*. Stuttgart: Hirzel.

Popper, K. 1935. *Logik der Forschung*. Wien: Springer. Tübingen: Mohr. Quoted after the Engl. transl.: *The Logic of Scientific Discovery* (1959). London and New York: Routledge Classics 2002.

Popper, K. 1949. Naturgesetze und theoretische Systeme. In: S. Moser (Ed.), *Gesetz und Wirklichkeit*. Innsbruck: Tirolia.

Popper, K. 1958. The Aim of Science. *Ratio* 1, 24–35.

Popper, K. 1974. Scientific Reduction and the Essential Incompleteness of all Knowledge. In: J. Ayala and Th. Dobzhansky (eds.), *Studies in the Philosophy of Biology*. Berkeley: University of California Press, 259–284.

Popper, K. 1975. The Rationality of Scientific Revolutions. In: R. Harré (Ed.), *Problems of Scientific Revolutions*. Oxford: Oxford University Press, 72–101.

Popper, K. 1979. *Truth, Rationality, and the Growth of Knowledge*. Frankfurt a. M.: Klostermann.

Reichenbach, H. 1938. *Experience and Prediction*. Chicago: Universit of Chicago Press.

Scheibe, E. 1976a. Gibt es Erklärungen von Theorien? *Allg. Zeitschr. f. Philosophie* 1, 26–45. Engl. transl.: Are there Explanations of Theories? In: D. E. Christensen et al. (Eds.), *Contemporary German Philosophy*, Vol. 3. London: Penn State University Press 1983, 141–158. Repr. in: Scheibe 2001, 324–338.

Scheibe, E. 1982a. A Comparison of Two Recent Views on Theories. *Metamedicine* 3, 233–253 [slightly extended version of 1981c]. Repr. in: Scheibe 2001, 175–194.

Scheibe, E. 1984a. Zur Rehabilitierung des Rekonstruktionismus. In: H. Schnädelbach (Ed.), *Rationalität. Philosophische Beiträge*. Frankfurt a. M.: Suhrkamp, 94-116. Engl. transl.: Towards a Rehabilitation of Reconstructionism. In: Scheibe 2001, 195–211.

Scheibe, E. 1987a. Ganzheitsaspekte in Philosophie und Wissenschaft. In: *Jahrbuch der Akademie der Wissenschaften*. Göttingen: Vandenhoeck & Ruprecht, 39–54. Engl. transl.: Aspects of Wholeness in Science and Philosophy. In: S. N. and Y. B. Tripathi (Eds.), *Medical Systems with a Holistic Approach*. Varanasi 1993, 13–29. Repr. in: Scheibe 2001, 23–35.

Scheibe, E. 1987un. *Explanation, Reduction, Progress. Investigations into the Unity of Physics*. A Seminar given at the Philosophical Department of the University of California, Irvine 1987. Unpublished lecture. https://doi.org/10.17877/DE290R-22744.

Scheibe, E. 1988a. *Stud. in History and Philosophy of Science* 19, 141–159. Reprined in: Scheibe 2001, 90–107. German version: Der Fortschritt der Physik im Selbstverständnis der Physiker. In: W. Muschik und E. Scheibe (Eds.), *Philosophie Physik Wissenschaftsgeschichte*. Berlin: TU Berlin, 1988, 165–183. German version reprinted in: M. Kroß (Ed.), *Vom Verständnis der Natur.* Einstein Forum, Jahrbuch 2000. Berlin: Akademie Verlag, 127–147.

Scheibe, E. 1988b. Paul Feyerabend und die rationalen Rekonstruktionen. In: P. Hoyningen-Huene und G. Hirsch (Eds.), *Wozu Wissenschaftsphilosophie? Positionen und Fragen zur gegenwärtigen Wissenschaftsphilosophie*. Berlin: de Gruyter, 149–171. Engl. transl.: Paul Feyerabend and Rational Reconstructions. In: Scheibe 2001, 212–227.

Scheibe, E. 1993a. Heisenbergs Begriff der abgeschlossenen Theorie. In: B. Geyer et al. (Eds.), *Werner Heisenberg. Physiker und Philosoph*. Heidelberg: Spektrum, 251-257. Engl. transl.: Heisenberg's Conception of a Closed Theory. In: Scheibe 2001, 136–141.

Scheibe, E. 1993c. A New Theory of Reduction in Physics. In: J. Earman et al. (Eds.), *Philosophical Problems of the Internal and External Worlds: Essays on the Philosophy of A. Grünbaum*. Pittsburgh: University of Pittsburgh Press, 249–271. Repr. in: Scheibe 2001, 352–368.

Scheibe, E. 1994a. Zwischen Rationalismus und Empirismus: Der Weg der Physik. In: F. Fulda and R. -P. Horstmann (Eds.), *Vernunftbegriffe der Moderne. Stuttgarter Hegel-Kongress 1993*. Stuttgart: Klett-Cotta, 73–95. Engl. transl.: Between Rationalism and Empiricism: The Path of Physics In: Scheibe 2001, 69–86.

Scheibe, E. 1995a. The Rationality of Reductionism. In: R. Zwilling (Ed.), *Natural Sciences and Human Thought*. Berlin: Springer, 101–109. Repr. in: Scheibe 2001, 369–377.

Scheibe, E. 2001. *Between Rationalism and Empiricism. Selected Papers in the Philosophy of Physics*. Ed. by B. Falkenburg. Berlin: Springer.

Sklar, L. 1967. Types of Inter-theoretic Reduction. *Brit. J. for the Phil. of Sci.* 18, 109–124.

Sneed, J. 1971. *The Logical Structure of Mathematical Physics*. Dordrecht: Reidel.

Spector, M. 1978. *Concepts of Reduction in Physical Science*. Philadelphia: Temple University Press.

Stegmüller, W. 1976. *The Structure and Dynamics of Theories*. Berlin: Springer.

Study, E. 1914. *Die realistische Weltansicht und die Lehre vom Raum*. Braunschweig: Vieweg.

Weinberg, S. 1987. Newtonianism, Reductionism and the Art of Congressional Testimony. *Nature* 330, 433–437.

Weinberg, S. 1992. *Dreams of a Final Theory*. New York: Vintage Books.

Weizsäcker, C. F. v. 1980. *The Unity of Nature*. New York: Farrar, Straus and Giroux 1980.

Yoshida, R. M. 1977. *Reduction in the Physical Sciences*. Halifax, NS: Dalhousie University Press.

Chapter 2
Physical Theories

What we are connecting with each other in reductions are theories. A sufficiently precise notion of theory therefore belongs in our foundations. In terms of the concept of theory, there has recently been a change of opinion regarding an elementary point about which a position should be taken here right from the beginning. The analytic philosophy of our century initially had a philosophy of language phase. Dummett expressed this in summary as follows (Dummett 1978, 442):

> For Frege, as for all subsequent analytical philosophers, the philosophy of language is the foundation of all other philosophy because it is only by the analysis of language that we can analyze thought . . . If the philosopher attempts, . . ., to strip thought of its linguistic clothing and penetrate to its pure naked essence, he will merely succeed in confusing the thought itself with the subjective inner accompaniments of thinking.

In the philosophy of science of logical empiricism, through Carnap, one of the two main directions of modern language philosophy, the so-called ideal language direction, came to the fore. With great exclusivity, it turned to the *syntax* of logical artificial languages which came into question in the reformulation of scientific theories. A *formalism* became an explicit part of theory and the notion of formal language became part of the conception of a theory. However, in the 40s, Carnap introduced Tarskian *semantics* into the philosophy of science and, accordingly, a particular formalism as well as its *interpretation* was accepted as a legitimate part of a scientific theory. His approach thus could be described as, at first, syntactic and then, later (in the sense of extension), semantic.

In the course of the, in some respects, more spontaneous than deliberate, post-positivist 'reforms' of the philosophy of science, one degenerated in regard to this point to the other extreme, and has sworn off any linguistic objectification, and thus any explication of the language as a component of a theory. This did not mean turning away from the use of formal means as such—nor a return to psychologism. It was just that the formal means migrated to the *meta-theory*. The previous objectification was partly regarded very harshly as a total philosophical mistake as, for example, by van Fraassen (1986, 56) when he writes:

E. Scheibe, *The Reduction of Physical Theories*, Fundamental Theories of Physics 207,
https://doi.org/10.1007/978-3-662-65000-4_2

> Perhaps the worst consequence of the syntactic approach was the way it focused attention on philosophically irrelevant technical questions. It is hard not to conclude that those discussions of axiomatizability in restricted vocabularies, 'theoretical terms', Craig's theorem, 'reduction sentences', 'empirical languages', Ramsey and Carnap sentences, were one and all off the mark—solutions to purely self-generated problems ... The main lesson of twentieth century philosophy of science may well be this: no concept which is essentially language dependent has any philosophical importance at all.

The movement against the older 'syntactic approach', as van Fraassen calls it, was given the name 'semantic approach' (cf. Suppe 1974, 221–230). But both descriptions are misleading because the logical-empiricist approach, which was initially in fact syntactic, became semantic as time went by, and the new direction is strictly understood as certainly not semantic, nor wishing to be. The newer so-called 'structuralism' has been spared this terminological misguidance. But, in fact, this direction is also expressly in line with the new course (recently, cf. Balzer et al. 1987). In the words of (Sneed 1976, 144, Note 2), it is that

> ... the way of talking about scientific theories I am going to describe invites us to look at sets of 'models' for these theories rather than the linguistic entities employed to characterize these models...

One reason or another can certainly be given for the expulsion of the linguistic element from the notion of theory and the rejection of any linguistic explication within the philosophical context in favor of a renewed direct orientation 'toward things'. But given the radical way in which it is expressed in van Fraassen's words and is also represented in structuralism, it is one of the typical philosophical exaggerations that must always be taken back soon. Its only goal is usually quickly achieved: the wakening of the perpetrators and victims of the respective preceding exaggeration from the 'dogmatic slumber' into which they had meanwhile fallen. It is certainly not clear why the realms of what is philosophically interesting and uninteresting should coincide just with those language-independent or -dependent phenomena, as is now claimed, or vice-versa, as was the fashion before. A priori, it is far likelier that there are interesting philosophical questions that concern language and others that do not concern it.

About what the more specific point concerns, namely, to dispense with the object-language within the concept of theory, it is abundantly clear, and the words of Sneed also unequivocally demonstrate, that this step does not mean the acceptance, but rather the abandonment of semantic terms in the strict sense: These notions are in fact meant to *combine* formal linguistic elements with corresponding elements of content. But, if the former are gone, there is already nothing more to be combined. Anyone who wants to eliminate the means of expression at the object level cannot label 'semantic' what he still wants to do without greatly misrepresenting his position from the outset. Carnap has already developed the appropriate terminology in his semantics, and, in that, one would have to say that the so-called 'semantic approach' is limited to *absolute concepts* (cf. Carnap 1959, Sects. 10 and 17). The most important example of an absolute concept is the concept of *structure*. This concept emerges from the *de facto* semantic notion of model when omitting from the model that for which it is a model: the formal theory and its language. (Sometimes, however, the latter is maintained, cf. Shoenfield 1967, 2.5). So what Sneed really wants to say in the above quote is this: One replaced a formal theory initially by the class of its models, and then talks in a second step merely of classes of structures, without regard to their formal origin. This is a grand (in fact, more thoroughgoing) generalization, but it has nothing to do with semantics, as long as one does not yet return to formalizations at the object level.

It is just this return that is intended in this book, and our undertaking could, therefore, rightly be called a 'semantic approach.' It is related, if not identical, to the approach of Ludwig (1990). Here, we must forsake the pros and cons of pure 'absolutism'. Our method is to be *as general as necessary and as specific as possible* in the physically relevant matters of principle. At least the second half of this directive was strikingly neglected by the direction in question. No single physical circumstance has yet been exhibited which would force or at least suggest this generalization of the concept of theory. On the other hand, we will see that the inclusion of certain sets of models of formal theories is, in fact, a great advantage when it comes to distinguishing and clarifying the notion of approximate reduction. (Compare this approach to that of Da Costa and Doria 1992.)

2.1 Analysis of an Example

Every physicist knows how to describe *in general* the motion of a particle in a force field in Newtonian mechanics. It is done via the equations

$$md^2x_i/dt^2 = \alpha f_i(\mathbf{x}) \quad (i = 1, 2, 3) , \tag{2.1}$$

where m is the mass, $x_i(t)$ the position coordinate of the particle at time t, α a further parameter dependent on the particle (e.g., the charge), and the $f_i(\mathbf{x})$ (most precisely, $\alpha f_i(\mathbf{x})$) being the force components present at the position $\mathbf{x}$. In physical practice, it is essentially specified in this way, and one can study the motions once the force field is known. Of course, our equation of motion in itself is just a piece of mathematics. It obtains physical meaning only against a certain background, which the physicist will consider as soon as he is faced with the equation and which is given by the concepts of space, time, mass, and force in the present case. Behind the equation lies a *physical theory* which typically contains mathematical concepts in addition to physical concepts. A physical theory characterizes a class of physical systems, the latter being regarded as structures that are, in general, composed of physical and mathematical elements. This must now be explained in more detail, using the example.

Physicists like to denote the objects with which they deal as *physical systems*. This denomination primarily indicates that their objects are already subject to a scientific approach, specifically as objects of physics, understood and treated as such, in contrast to the commonplace things we encounter. A pendulum, a magnetic field, a capacitor, a crystal, a hydrogen atom, the sun, and the solar system are typical objects of physics which, in part (and in the predominant part, as some emphasize), do not occur in our ordinary experience but first must be established or made perceptible with more or less great technical effort. Even here, therefore, the question arises as to whether the views of physics expressed by this denomination are capable of a reconstruction capturing its characteristic features. The answer given in this book is: A genuine characterization in the sense of necessary and sufficient conditions of something being a physical system will not be found (cf. Scheibe 1979). As a condition which is only necessary, the conception of a physical system as a *structure* can be considered.

Structures are familiar to us in ordinary life mainly *geometrical* structures. Here, the special and already nearly scientific perspective from which we reconstruct a thing as a geometric structure is denoted by the concept of *form*: Lines and planes, circles and spheres, parabolas and cones are objectified forms of ordinary geometry, in terms of which we can identify the shape of things we encounter empirically. Of course, only a few, and even these only through some (often considerable) degree of idealization: A tabletop is never exactly circular or rectangular, a rail never exactly straight, and an ice rink not flat in the sense of geometry. And, of the tree out my window, I do not know the specific shape. Nevertheless, geometrical concepts have proven useful for the spatial description of the world of things. They define for us types of structures, and we usually assume that all things in space have a particular shape, even if these are too complicated to determine in most cases and the corresponding concepts are lacking that express the result of such an identification.

In what sense are figures now structures? First of all, they are only so in a relative sense: Lines and planes, parabolas and circles are in modern thought essentially figures in *space*, and we usually understand their meaning in terms of a particular *theory* of space, *Euclidean geometry*. If we are asked what a circle is, we say that it is all of the points having the same distance from a given point of a given plane. So we understand the circle—so defined—as a set of points among a larger set of points, the plane, and we refer in our definition to a further structure of space which is indicated by the words "distance between two points" and which we assume satisfies the laws of Euclidean geometry. A circle, a straight line, and so on, therefore, do not exist by themselves in an obvious sense, but they are the circle, line, and so on in Euclidean space. Only with this—with space as a set of points—supplied with a Euclidean distance, do we appear to have found something that accomplishes the specification of individual figures: We are not disposed, in the sense in which we say "a circle is just a circle in space," to think of Euclidean space as well that it is only a space in... Of course, that does not mean that Euclidean space is in some absolute sense an ontologically independent structure. It is just that physics has taught us that it is not. But it means that, after all, we have a remarkably closed theory of space of a sort that we couldn't provide for a figure *in* space.

But we are not yet dealing with this theory at the moment. Right now, the concern is answering the following question: In what sense is Euclidean space a structure, and in what sense, then, are the figures within it structures as well. Now, space appeared to us as a set of points provided with a distance relation—say a set of points S and a distance relation d. We deal thus with the pair

$$\langle S; d \rangle \,. \tag{2.2}$$

For d, the question of what d is relative to S can be answered—provided, for numerically specifying distances, we take in addition the set $\mathbb{R}$ of real numbers. Then, we can say in particular that d is a relationship between any two points $P_1, P_2 \in S$ and a number $\delta \in \mathbb{R}$ that exists precisely if P_1 is in a distance δ from P_2. With Pow(X) the power set, i.e., the set of subsets of X, and with $Y \times Z$ being the Cartesian product of Y and Z, i.e., the set of pairs $\langle y, z \rangle$ with $y \in Y$ and $z \in Z$, we then have

$$d \in \mathrm{Pow}(S \times S \times \mathbb{R}) \tag{2.3}$$

as the answer to the question of what d is in relation to S. And, for all the products of figures in space listed above in relation to S, it would be valid to consider them subsets of S. Here, we thus had

$$M \in \mathrm{Pow}(S) \tag{2.4}$$

for every such M. But, as far as the elements of S itself, $\langle S; d \rangle$ does not tell us what their natures within our structure are. Euclid was already acting in vain when he tried to define what points are by saying that a point is that which has no parts.

Thus, we say of Euclidean space that it is a structure insofar as it is a pair (2.2) of quantities such that the nature of the elements of d is known according to (2.3), while the nature of the elements of S within the structure remains open. It can only be known to some extent from the outside. In contrast, there is lack of independence of those figures in space due to the fact that they are—much like d already—*typified*[1] by S, as one says. One can, through distinction of figures $M_1, M_2, \ldots$, expand the space $\langle S; d \rangle$ to a structure

$$\langle S; d, M_1, \ldots, M_l \rangle \,, \tag{2.5}$$

and it is such an 'absolute' structure in which the M_i appear properly as 'relative' structures: circles are circles in Euclidean space, not circles in themselves.

The reconstruction of space as a structure in the sense indicated, of course, still yields no information whatsoever regarding the fact that we have considered this space *Euclidean*. This must be enunciated only through statements about $\langle S; d \rangle$—through the laws of Euclidean geometry. For this purpose, there is first the synthetic method, originating with Euclid himself, according to which something is said *directly* concerning the distance function d. A typical example would be the triangle inequality

[1] The neologism *typify (typisieren)* is based on the term *typification (Typisierung)* introduced by (Bourbaki 1968, 261) (also in French: Bourbaki 1970, E IV 3) and adopted by Scheibe (1979, 1982a, 1986a) to express the structure of physical theories (see Scheibe 2001, 164, 181, and 540).

$$\text{for all } x, y, z \in S : d(x, y) + d(y, z) \geq d(x, z) , \tag{2.6}$$

where it is already assumed of d that it is a function

$$d : S^2 \to \mathbb{R} . \tag{2.7}$$

If one wants to obtain an axiomatic theory of this type, there is much to do. We arrive at the goal more quickly by the analytical method with which an *indirect* statement is made about d. One requires the existence of a global coordinate system ϕ on S in which

$$d(x, y) = \left\{ \sum_i (y_i^\phi - x_i^\phi)^2 \right\}^{1/2} , \tag{2.8}$$

where x_i^ϕ and y_i^ϕ are the coordinates of x and y, respectively, in ϕ. It is obvious that propositions of the sort of triangle inequality are physically more transparent than the propositions of (2.8). The problem of replacement of such indirect statements by equivalent direct ones is one that we will have to deal with repeatedly. But, for the moment, it suffices that Eq. (2.8) appears to bring about the tour de force of a complete characterization of the Euclidean geometry.

The geometric structures in the usual sense are not just examples which are especially easily accessible to laymen but are to be taken seriously as fragments of physical systems: As these are in space, they necessarily, in one way or another, have a geometrical aspect. With the assumptions (2.2, 2.3 and 2.8), we are thus on the way toward a theory centered around Eq. (2.1). The physical content is, however, further enriched if we also consider *kinematic structures*. In order to do so, we introduce *time* as a structure in addition to space. For, just as circles are only circles in space, a kinematic structure, e.g., the motion of a planet, is a structure in space *and time*. Time is treated analogously to space in physics: One regards it as a set T of time points, which is also provided with a distance structure e and, additionally, with an ordering structure e_1. In the absolute sense, we hence deal here with a structure

$$\langle T; e, e_1 \rangle . \tag{2.9}$$

The distance e is the duration extending between two points in time. The typification is, therefore, like that in space

$$e \in \text{Pow}(T \times T \times \mathbb{R}) . \tag{2.10}$$

We express that one point in time is earlier than another through the ordering e_1. Therefore, we have the typification

$$e_1 \in \text{Pow}(T \times T) . \tag{2.11}$$

Again, as already with space, the main question now arises of which assumptions to make regarding the temporal structure (2.9), and, again, we want to answer in the simplest way using coordinate systems: We demand the existence of a (global) coordinate system ϕ on T such that

$$e(t_1, t_2) = |t_1^\phi - t_2^\phi| \, , \; \langle t_1, t_2 \rangle \in e_1 \leftrightarrow t_1^\phi < t_2^\phi \, , \tag{2.12}$$

wherein e is again written as a function and t_i^ϕ are the coordinates of $t_i \in T$ in ϕ. That is already the entire time axiomatics.

The list of our *dramatis personae* is now continued with the field explicitly appearing in Eq. (2.1) with the components $f_i(\mathbf{x})$ at position $\mathbf{x}$. A (time-independent) force field is understood in Newtonian mechanics to be a function that assigns to each point in space a force which is just the force acting on a particle that itself has that point in space as its position. So if we want to continue further in the sense of the foregoing remarks, we need to introduce a set K of possible forces, which are then somehow distributed in a force field over space. The force field, or rather, particle-independent field, which is independent of α as in Eq. (2.1), had the typification

$$f \in \mathrm{Pow}(S \times K) \, , \tag{2.13}$$

while the elements of K remain untypified as possible field strengths, and thus ontologically independent entities, as do the space and time points. But that's not all. K is connected with S by the fact that a (global) coordinate system B_ϕ on K uniquely corresponds to every Euclidean coordinate system ϕ on S, which is subject to (2.8). Thus, we have here a further structural element B, a relation with the already complicated typification

$$B \in \mathrm{Pow}\{\mathrm{Pow}(S \times \mathbb{R}^3) \times \mathrm{Pow}(K \times \mathbb{R}^3)\} \, . \tag{2.14}$$

B also has the important characteristic that a coordinate transformation

$$x_i^\psi = \sum_j a_{ij} x_j^\phi + a_i \, , \tag{2.15}$$

with orthogonal a_{ij} on S, corresponds to the coordinate transformation

$$f_i^{B\psi} = \sum_j a_{ij} f_j^{B\phi} \tag{2.16}$$

of the force components. As a shorthand, one often calls the $f_i^{B_\phi}$ the force components in ψ (in place of B_ϕ). The distinction of a class of coordinate systems on K finally permits the introduction of a vector-space structure and a modulus function on K in the usual manner.

The previously introduced structure, consisting of space and time, and the (α-independent) field, each with associated accessories, forms the *core* of our theory. By the core of a theory, we mean the portion of the physical systems treated by the theory which is common to all systems that obey the theory. The decision of what we want to belong to the core of a theory is our choice in some cases and not in others. Space and time (including their metrical structure) in Newtonian physics, viewed globally, are given only *once*, and when we are comfortable with mechanics and optics as different theories, for example, then this difference does not extend yet to space and time, which are the same in both theories. In contrast, it is in the present case somewhat arbitrary whether one counts the field as also belonging to the core of the theory or not. We want to express that we consider the field as the fixed environment of the particle, of which it is independent, by which it undergoes no reaction at all and, accordingly, that it does not belong to the actual object of the theory and is not subjected to a field equation, etc. This decision is completely independent of the question of *which* field is involved. Although we imagine that the field for each theory, as we now delimit it, is different, we need not address that question. As long as we treat all these theories together, we are not *able* to give the answer at all.

From the fixed core of our theory, we finally differentiate its *variable part*, which in the present case is given by the mass m, the parameter α, and the motion $\mathbf{x(t)}$ of the particle, with the typification

$$m, \alpha \in \mathbb{R}^+, \quad \mathbf{x} \in \mathrm{Pow}(T \times S) \,. \tag{2.17}$$

Here, again, as already assumed for the spatial and temporal distances, we assume a standard scale for m and α, while the motion is first directly related to space and time. But, when it comes to the question of which law of motion m, α, and $\mathbf{x}$ satisfy, then it is easiest to demand again that, in a spatio-temporal Euclidean coordinate system, the mathematical representations $x_i(t)$ of $\mathbf{x}$, and $f_i(x)$ of $\mathbf{f}$ together with m and α satisfy the Eqs. (2.1). However, the $x_i(t)$ do not go directly into these equations, but rather their second derivatives—the accelerations. Formally viewed, in common usage, not only are the basic concepts of a theory used for the convenient formulation of statements, but also concepts which are defined in terms of them. But as regards content, we also have the problem of the sense of 'definition' that should be used. What *is* the acceleration of a motion and even its velocity?

The problem comes, in essence, down to the question of whether we should expect (or whether the adoption is permitted at all) for the magnitudes of velocity and acceleration to be certain well-defined constructions out of the respective motion, or whether it is better to give them a distinct status *sui generis*. The answer is not self-evident. In fact, one often obtains the speed and acceleration as a local function of time by differentiation. But one sometimes also sees in physics velocity fields in the abstract (e.g. in continuum mechanics), and only then sets up a differential equation for motions of the given velocities. On the other hand, one should remember that force-like tensorial quantities are basically identified with the canonical tensors derived from the differential geometry of the underlying manifold in the general theory of relativity, without even posing the question of substantial identity. The question of how one has to proceed here is no way clearly answered in physics. We want to leave the matter in suspense, for at the moment we do not have the necessary resources available, and, in the present case, emphasize the following.

By (2.8) and (2.12), a set F of coordinate systems on space and time (combined) is distinguished, of which any two such $\phi, \psi \in F$ arise from each other via a coordinate transformation

$$x_i^\psi = \sum_k a_{ik} x_k^\phi + a_i \, , \quad t^\psi = t^\phi + a \tag{2.18}$$

(with $a_{ik}a_{lk} = \delta_{il}$). These coordinate systems can be used to represent all structural elements typified on space and time. So the motion $\mathbf{x}$, which itself also possesses the coordinate-free typification, (2.17) in a coordinate system $\phi, \psi \in F$, is represented by the function

$$\mathbf{x}^\phi : \mathbb{R} \to \mathbb{R}^3, \quad \mathbf{x}^\phi(t^\phi) = \phi(\mathbf{x}(\phi^{-1}(t^\phi))) \, , \tag{2.19}$$

and two representations result from one another via the transformation

$$\mathbf{x}^\psi(t^\psi) = \psi\phi^{-1}(\mathbf{x}^\phi(\phi\psi^{-1}(t^\psi)) \, . \tag{2.20}$$

We can condense all these representations in a structure with the typification

$$\mathbf{x} \in \text{Pow}\{\text{Pow}(S \times \mathbb{R}^3) \times \text{Pow}(\mathbb{R} \times \mathbb{R}^3)\} \; : \tag{2.21}$$

Every coordinate system $\phi \in F$ is assigned by a function $\mathbf{x}^\phi$ according to (2.19). This summary is highly redundant, something expressed by the various transformations (2.20).

Coordinate representations of type (2.18) can now be used even if they are based on *no* coordinate-free typification such as (2.17). This approach is somewhat neutral with respect to the question of the nature of the quantities represented. Thus, this results in the acceleration in the coordinate system $\phi \in F$ defined via the second derivative of $\mathbf{x}^\phi$ by a function

$$(d^2\mathbf{x}/dt^2)^\phi : \mathbb{R} \to \mathbb{R}^3 \tag{2.22}$$

with the transformation law

$$(d^2x_i/dt^2)^\psi(t^\psi) = \sum_k a_{ik}(d^2x_k/dt^2)^\phi(t^\psi - a) \tag{2.23}$$

and a summarization as in (2.21). One is, in this way, guided to this summarization by the above coordinate-independent definition of acceleration—acceleration in each coordinate system—and the question of what accelerations 'in themselves' are can remain open for the time being.

We have expanded the analytical method of geometry to physics straightaway here in the introductory case study because it is the most typical method and does seem to be unavoidable in many cases. We can see this clearly when we look back and realize how we have treated the distance function d and the mass m. In all cases, we have tacitly assumed that it is about real-valued functions or constants, respectively. Strictly speaking, we brought the real numbers only into play by distinguishing a 1-dimensional coordinate system on the respective range of values: Spatial and temporal distances and masses were reproduced *by means of* real numbers. And, strictly speaking, the theory does not uniquely determine these coordinate systems: The choice of the unit is arbitrary. This arbitrariness will only be limited again by Eq. (2.1). In any case, we should *conceptually* distinguish the physical magnitudes and the real numbers, even if the scale is unique from the physical point of view. And that would mean that we would have to introduce, beyond S, T, and K, additional basic sets, with the respective physical magnitudes (distances, masses, etc.) as elements. Especially when the scale is determined uniquely (or almost uniquely), however, working only with the real numbers as representatives in all cases means a great simplification.

What is the result of the above analysis of the dynamic theory (2.1)? It treats the motion of a particle in a force field in space as a function of time and essentially requires the use of the Newtonian equation of motion (2.1). We have reconstructed the underlying physical system—the world of this theory—as a structure that is composed of physical and mathematical sets.[2] The basic sets whose elements have to be interpreted 'from the outside' were S, T, and K, as well as the mathematical set $\mathbb{R}$. The latter appeared directly as a set of values for spatial and temporal distances, as well as for masses. All other quantities that occur as structural elements, such as d, e, etc., could be typified through the basic sets. A fragment of this structure was firmly connected as a core to the theory—only m, a, and $\mathbf{x}$ remained open under the axioms of the theory. Finally, for specifying the axioms, we have used the analytical method: The claims were not directly made about the physical quantities, but rather about their mathematical representatives in a coordinate system. The structures distinguished in this way as a whole are *possible* physical systems in the sense of the theory. It should be noted that the axioms alone do not bring about this distinction. In particular, they have the property that each structure isomorphic to a model of the axioms is also a model. A further restriction is only obtained by *directly* distinguishing a common fragment in all models, that is, the core of the theory, almost as if by an indication from God.

To the previously discovered pieces of a physical theory, yet another is now added which has to do with the application of the theory: its *domain of intended applications*. Our axiom system is not yet a proposition, but only a propositional form with variable parameters m, a, and $\mathbf{x}$ of each particle. Only from specifying

---

[2] To introduce physical sets is a peculiar aspect of Scheibe's approach. To do so, he complements the set theory of Zermelo and Fraenkel (ZF) with "Urelemente" in the sense of Zermelo (1908), which he identifies with individuals in the sense of (Field 1980, 9) (see below, next section, and (Scheibe 2001, 540).

them an assertion proper would result—namely, that the particle satisfies the axioms of the theory. In fact, according to the view taken here, a theory is a *concept* of physically possible systems which primarily enables us to make statements about individual systems. The universality of the theory in the usual sense enters only via determining one or the other domain of application A and claiming that *all* systems in A satisfy the axioms of the theory. With Σ as a summary of the axioms, so one gets something like:

$$\forall a: \text{ if } a \in A, \text{ then } \Sigma(S, d; \ldots; m^a, \alpha^a, \mathbf{x}^a), \tag{2.24}$$

where in Σ, first, all elements belonging to the core structure are listed, and m^a, α^a, $\mathbf{x}^a$ are the mass, characteristic parameters and motion of the particle a.

2.2 Logical and Set-Theoretical Background

From the perspective obtained in the first section, we look Janus-faced in two opposite directions. We have obtained the notion of a theory from the analysis of an example—more precisely, mere aspects of such a notion. So far, we know neither the logical foundations upon which it is based, nor even the first steps of its further development and empirical application. The following sections of this chapter are dedicated to some related questions.

First, as regards the logical foundations, this will include a generalized set theory. Among the reasons to choose set theory as an essential basis (next to first-order logic) range, first, that physics is highly mathematized, and that a *possible* basis for mathematics is just set theory, for example, the system ZFC of Zermelo/Fraenkel with the Axiom of Choice. Of course, one could also think of an intuitionistic foundation of mathematics, especially in the context of physics, or one could work directly with (classical) logical systems of higher order. Each of these possibilities would be enough to guarantee the component of mathematics used in physics. Because a system like ZFC delivers much more, its choice would be even less appropriate in the sense of adapting the mathematical apparatus to physics. But formal set theory is much more practical than the other possibilities just for our purpose of theory comparison, and we want to take this into consideration in an attempt that is still standing on shaky ground anyway.

However, the choice of a set theory as a basis has also other reasons that are less pragmatic. It is noteworthy that one happily associates mathematics with set theory today. Laymen are familiar with (at least finite) sets from contexts in which one does not immediately think of mathematics. The sets that we construct in everyday thought are collections of common items or persons. Its elements, at least, are just not mathematical objects, but rather those of experience as well as objects of physics. The fundamental relation of set theory, membership, has therefore, in general, relata that do not both belong to mathematics. The chair on which I sit is an element of the set of all chairs in my house in exactly the same sense in which the number 5 is element of the set of primes. So, it is not obvious at first why a set theory should exclude empirical and, in particular, material objects as potential elements of sets, or such objects themselves as sets, e.g., of molecules. Indeed, they were not excluded in the first axiom system of set theory presented by Zermelo (1908). Zermelo even admitted so-called *Urelemente*, individuals which in turn have no more elements, but yet are not identified with the empty set. Instead, they could be physical objects in the broad sense of the word, e.g., points of physical space as well as electrons and stars. One then, however, realized that actually certain *constructions* of sets from other sets constitute the mathematical character of the set concept and that, for the purposes of mathematics, the totality of sets constructed out of the empty set is sufficient. Since, for more historical reasons, mathematicians and mathematical logicians have become interested only in the later development of set theory, Zermelo's *Urelemente* fell from view and no longer appear in most treatments of the subject. For the reconstruction of a discipline like physics that is so closely associated with mathematics, however, a version of mathematical set theory extended by such individuals actually lends itself to use (cf. Field 1980, Chap. 1).[3] This is even more evident as we have shown in the previous section through an example that the concept of *structure* is very suitable for the reconstruction of a physical theory. This concept, however, is a typical set-theoretical one: A structure is a finite system of sets, some of which are typified by others. What this generally means (see the next section) can be said only by referring to two set-theoretic constructions: the Cartesian product and the formation of the power set. On the other hand, the structures most frequently found in physics are a mixture of mathematical and physical elements. How could one proceed, here, more appropriately than by treating the two types of elements with the *same* theory?

Our example in the previous section was a particle that moves in a force field. Thus, the total that we have before us is a structure composed primarily of physical objects but which also contains mathematical structural elements. This is evident already in the geometric part of the theory in two ways. Space is a set of physical elements. Of course, it is not about physical objects in the strict sense—it's not about bodies. But it is even less about mathematical entities. Because the points of physical space are, after all, idealizations of *potentially* materially marked spatial locations. The number triples introduced for the mathematical *description* of such a location are certainly something else. It is the actual point of such a description that, with a coordinate system, we get a theoretical *substitute* for the 'real' space. And, here, we have now one in which mathematical objects are incorporated into a physical structure: Beside physical space, we have the set of the Euclidean coordinate systems describing it.

[3] See also Scheibe 1986a, 1994a: "Field seems to think that if physical axiomatics is based on set theory at all, then we have to resort to a version allowing for individuals (= Urelemente in the sense of Zermelo)" (Scheibe 2001, 540, referring to Field 1980, 9)."Such a thing was developed by Zermelo already in 1908[10]. Zermelo introduced in a set-theoretical universe entities different from the empty set but, like it, having no elements. These individuals or *Urelemente*, as Zermelo called them, ... can be freely identified, for instance, with points of physical space or spacetime, with mass points, field strengths or other fundamental physical entities" (Scheibe 2001, 575–576; the note refers to Suppes 1960).

Yet, every single coordinate system is a mixed object because it precisely connects the points in space with the numerical triples: physical with mathematical entities.

With Euclidean space, we also have in other respects a structural hybrid ahead of us if we, as was done in Sect. 2.1, describe the distances between points *directly* by (positive) real numbers—directly, i.e., if we do not introduce the various physical possibilities for such distances specifically and then associate them via a (1-dimensional) scale with the real numbers, but simply take as a component of the Euclidean structure their numerical values *in place* of the physical distances between them. This participation of mathematical entities in a physical structure is more significant than the other, casual use of coordinate systems: The latter could be struck out without replacement, without a concern of losing 'physical substance', the former not—unless one explicitly puts the original physical distances back in their place. Then it would, here, again, regard the common use of coordinate systems. It is only that it has become common here, and in many similar cases, to carry along the spatial coordinate system *alongside* space but to take the numeric distance values as part of a Euclidean structure *in place* of the possible distances themselves. The reason is that, though the coordinate systems are determined only up to a Euclidean transformation, the numerical distance is determined up to a positive factor. Thus, the latter represents its 'archetypes' less arbitrarily than the former. The foregoing should make only plausible what set theory can contribute to the formulation of physics and how mathematical next to physical entities come into play. The formulas (2.2–2.8) and (2.9–2.12) of the previous section already summarize this motivation. Before we now raise these relationships systematically, a few words about logic are in order, as we need these even before any set theory (cf. Shoenfield 1967). According to the traditional view, logic is the science of inference. Reasoning is made possible because in the language of a theory, in addition to descriptive signs which serve the expression of genuine content directly (e.g. in set theory: x is an element of y), we also have logical signs—words such as "and", "not", "if-then", "for all" etc. These words mean something in the sense that they contribute to the truth or falsity of a proposition in which they occur, albeit in a less tangible way than descriptive words such as "...is at a distance of 3 cm ...", "...has a voltage of 2.7 V", etc. Now, if the truth of a proposition is guaranteed solely on the basis of the truth of certain propositions b, c, d, ...*together with* the importance of logical signs in all of these statements, then we say that a logically follows or could be inferred from the b, c, d, ... And, if the premisses are even allowed to be absent, we say that a is logically true. A famous logical rules of inference is the modus ponens

$$a, a \rightarrow c \vdash c\,, \tag{2.25}$$

where $\vdash$ is the inference sign and $\rightarrow$ stands for "if-then". A well-known logically true sentence is the tertium non datur:

$$a \vee \neg a \tag{2.26}$$

with $\vee$ for "or" and $\neg$ for "not." The advantage of this conception of logical consequence and logical truth is that it saves us the question of a genuine logical object. A logical proof is, thus, not correct and a logically true proposition is thus not true because they make an adequate claim about an object that could become incorrect or false by applying it to any other object, such as in mathematics wherein the proposition "5 is a prime number" would be wrong if we move on from 5 to 6. That which makes the proofs of logic correct, and a logically true proposition true, is not the object of the respective proposition, and it is not mentioned but rather used via the logical symbol.

Now, why are these seemingly insignificant relationships, why is logic important, perhaps indispensable in physics? The answer is: Because we draw conclusions in physics, for example, when we make predictions. The main focus of the physicist is indeed on the acquisition of acceptable premisses: certain empirical data and a physical law, the validity of which guarantee the validity of the prediction. But a prediction that was not inferred correctly might as well be wrong. The power of logical particles in reasoning can be made even clearer by the following illustration. We may think of a solution of the differential equation representing a physical law as a huge, infinite set of elementary propositions about a physical system. This catalog is logic-free: Using a portion thereof, we cannot readily infer a proposition of the catalog that is not already contained in this part. But, if we add the differential equation from a possibly very small part of the catalog, we *can* open up the whole rest. And that is essentially due to the fact that the differential equation has an internal logical structure (cf. Sect. 3.1).

Hence, we cannot do without a logic. But, given that set theory also belongs to our foundations, we get by with a first-order logic. The variables then only refer to the properties of the underlying object domain, and in addition to the rules and axioms of propositional logic, such as (2.25 and 2.26), we have the axioms of predicate logic, such as

$$\forall x . Fx \rightarrow Gx . \rightarrow (\forall Fx \rightarrow \forall Gx) \,, \tag{2.27}$$

with $\forall$ for "for all" and (possibly defined) predicates F and G. As elementary predicates, we have in set theory, in addition to the equality $=$, only the membership $\in$ and the empty set $\emptyset$ as a distinguished object. Moreover, we want to work with the fiction that with this language, which is formed as usual on this basis, about a realm of objects $\mathcal{L}$, which consists of sets and individuals and in which the $\in$ is interpreted as ordinary membership. The question of which realm of objects it regards remains open for the time being.

Since the usual axioms of set theory contain the arithmetic (thus it becomes just mathematically capable), it is essentially incomplete according to a famous theorem of Gödel and, therefore, one is left to choose which specific codification to make. Now, we want to connect with the hereby given, general freedom no special point and, therefore, limit ourselves, as already for logic, to only one example which makes clear the nature of the set-theoretic axioms (cf. Fraenkel et al. 1973, Ch. II, and Suppes 1960 as detailed treatments). In set theory, harmless assumptions combine with less harmless ones. In any case, we must keep the difference between sets and non-sets, i.e., individuals, in mind. Sets other than the empty set are those objects in $\mathcal{L}$ that have elements. We signify by $x, y, z, \ldots$ any objects in $\mathcal{L}$, but by $A, B, C, \ldots$ only sets, so that we have as a harmless axiom the axiom of extensionality

$$\forall x.x \in A \leftrightarrow x \in B. \rightarrow A = B\ , \tag{2.28}$$

which speaks for itself. The axiom of separation

$$\exists B \forall x.(x \in B) \leftrightarrow (x \in A \wedge \phi x).\ , \tag{2.29}$$

wherein ϕx is any propositional form, is only apparently harmless. If we omit the term $x \in A$ here, then with $\phi x \equiv \neg(x \in x)$ Russell's paradox follows. The belief that, for a properly formed propositional form ϕx, the *set* of x always exists is erroneous. The restriction to an expected given set A in (2.29) is, thus, already a somewhat artificial precaution. An axiom not threatened by paradox but rather of fulminant effect when infinite sets are available, is, as a third example, the power set axiom

$$\exists B.\forall C.C \in B \leftrightarrow C \subseteq A..\ , \tag{2.30}$$

that assigns to every set A the *set* of its subsets.

However, *one* of the usual set-theoretic axioms, whose acceptance often is referred to as a mere matter of taste, has a point in connection with our aim of physical application. It is the *axiom of foundation*

$$A \neq \emptyset \rightarrow \exists x.x \in A \wedge \forall y.y \in x \rightarrow \neg y \in A..\ , \tag{2.31}$$

i.e., that a non-empty set has an element with which it has no element in common. From this axiom (and the others), it follows that every chain[4]

$$x \ni y \ni z \ni \ldots \tag{2.32}$$

terminates after finitely many steps (cf. Fraenkel et al. 1973, 90). Sets thus become large not in depth, but rather in width. This axiom now allows us the following tripartite division of the objects of $\mathcal{L}$. First, we have those sets A for all chains (2.32) which end with $\emptyset$. A and all its elements have this property, as well as all elements of B for a given $B \in A$, etc. A is, thus, in its "substance" built up entirely from the empty set and the sets obtained by set-theoretic construction. Such a set is, therefore, an excellent candidate for the status of a purely *mathematical object*. Let us go over, on the other hand, to those objects of $\mathcal{L}$ in which all chains end with an individual. These include the elements, all sets of individuals, set of such sets and individuals, etc. So, if the individuals are taken to be by their nature physical objects, we clearly obtain in this second way the most common *physical objects* of $\mathcal{L}$. What remains are those objects whose chains (2.32) sometimes end in $\emptyset$ and, at other times, end in an

[4] Our translation of *Kette* by *chain* follows (Scheibe 2001, 541, Note 11).

individual: *mixed objects*. Of course, it is no longer the case that they also transmit their characteristics to their elements, and their elements, etc. Nevertheless, we have also, and particularly, candidates from physics for them. We want to illustrate this again with an example.

Above, we treated the Euclidean space as a physical entity, but one which has mathematical 'accessories' and, therefore, strictly speaking, is a mixed object. In its case, great pains have again been taken to remove the coordinate systems and the numerical distances, the mathematical element which only became so popular via physics. In fact, geometry was since ancient times just a synthetic and number-free doctrine of space until the 17th century, when the idea of analytic geometry arose (see Sect. 2.3). But let us accept the distance function d, as it was conceived in Eq. (2.3) in the previous section. Then $d \subseteq S^2 \times \mathbb{R}$ holds. The elements of S, the points in space, are individuals, hence physical, the elements of $\mathbb{R}$, as real numbers, are mathematical. The elements of S^2 are also physical. For they are pairs of physical objects. But none of the elements of $S^2 \times \mathbb{R}$ are physical. With the usual definition

$$\langle x, y \rangle \equiv \{\{x, y\}, x\} \tag{2.33}$$

for pair formation and with $x \in S^2 \times \mathbb{R}$, $y \in \mathbb{R}$, we have the chains

$$\{\{x, y\}, x\} \ni x, \ \{\{x, y\}, x\} \ni \{x, y\} \ni y \ , \tag{2.34}$$

with mathematical y and physical x, thus, a mixture that, of course, holds for all elements of d. The tripartite division of $\mathcal{L}$ induces a classification of the propositions of our set-theoretic language. We first isolate the mathematical language. Here, we get the clearest picture if we describe the situation by the claim that ordinary set theory, which concerns from the outset only mathematical sets, has a *conservative interpretation* in our set theory with individuals, that is, for example, the system ZF in ZFU.[5] If we denote the membership in both formalisms with $\in$ and the variables in ZF with $a, b, c, \ldots$, then the interpretation of a sentence α in the language of ZF by a sentence α' in the language of ZFU is obtained by replacing in each occurrence of quantifiers

$$\begin{aligned} \exists a \ldots \text{through } \exists x.ma(x) \wedge \ldots \\ \forall a \ldots \text{through } \forall x.ma(x) \wedge \Rightarrow \ldots \ , \end{aligned} \tag{2.35}$$

where $ma(x)$ states, in ZFU, that x is mathematical. So, for the axiom of infinity of ZF

$$\exists a.\emptyset \in a \wedge \forall b.b \in a \to \{b\} \in a.. \ , \tag{2.36}$$

we get the translation

$$\exists x.ma(x) \wedge \emptyset \in x \wedge \forall y.ma(y) \to (y \in x \to \{y\} \in x).. \ , \tag{2.37}$$

[5] ZFU is ZF with individuals ("Urelemente").

i.e., there is an infinite *mathematical* set. Generally, it emerges for this translation of α to α' that

$$\vdash_{ZF} \alpha \Leftrightarrow \vdash_{ZFU} \alpha' \, . \tag{2.38}$$

From left to right, it states that all theorems valid in mathematical set theory ZF *under restriction to mathematical sets* also hold in ZFU. One proves this first for the axioms and, then, uses the interpretation theorem for any remaining theorems (cf. Shoenfield 1967, 61–63). In the direction from right to left, it is stated that the interpretation is conservative, i.e., in ZFU no *more* theorems can be proved about mathematical sets than there already are in ZF.

The result (2.38) can be briefly, if somewhat imprecisely, expressed as that the truths about mathematical entities, and only those, are mathematical truths. The situation changes when we come now to those statements about $\mathcal{L}$ which concern individuals and, thereby, according to our arrangement, physical objects. Then, as with $ma(x)$, we have the predicate $ph(x)$ as "x is a physical object" at our disposal, with which we can explain what a statement *only* about physical objects is, using the schema (2.35). But, here, it turns out that statements which we *intuitively* judge as physical statements are not always also statements *only* about physical objects. *Most* statements of physics, whether law-like or contingent, are not. Indeed, the most obvious reconstruction of physical statements within our set theory would result in statements about physical *and* mathematical objects. One will most easily reconstruct them as statements about mixed sets, more precisely, structures (see Sect. 2.3). But that is just a special case of the general situation formulated just now. The (contingent) statement that the number of planets is 9 assigns to the set of planets, a physical set, the mathematical set 9 as its cardinal number, in our reconstruction. Although the number 9 appears in it, this is a physical statement and, vice versa, although it is a physical statement, the number 9 appears in it. The former is the case because it is 'really' about the set of planets, the latter because here you *want* to bring the number 9 into play as a count: It is about answering the question of what number belongs to an empirically given multitude. Also, at the level of law, it is considered normal that we make physical statements by bringing mathematical objects into play. Here, particular attention is paid to the special role of the empty set. On the one hand, it is a mathematical set and, on the other, one can hardly do without it in physical statements—it inevitably appears and may do so in the most benign way. So the important break thus lies between the statements about mathematical objects on the one hand and non-mathematical, i.e., physical or mixed objects, on the other.

Of course, there is also the case in which we make only statements about physical objects (or at least basically only about them), and some consider it an ideal to bring physics completely to a such formulation (cf. Field 1980). Indeed, at least we do not *want* in any case of mixed statements to have mathematical objects in them. We will immediately examine the case of Euclidean geometry more closely. And of particular interest today are the attempts to formulate quantum mechanics in purely physical terms (cf. Ludwig 1985). Once we have given the problem a precise formulation, we will continue to pursue it only marginally. Furthermore, we will consider the approach to the distinction of physical and mathematical objects and statements outlined in this section as mandatory background for what follows, but we will not really explicate it in all cases. To carry it out in detail would also affect the reductions, and precisely therefore go far beyond the scope of our actual program, which is to get the reduction of physical theories done without too much technical effort.

2.3 Physical Theories

After the thorough preparations of the two previous sections, we can now establish the most important basis for the later treatment of theory reductions: the concept of *physical theory*. However, there can be no question of truly narrowing this concept to provide something like a definition. It must suffice for us to introduce some important components of the theory concept, and the total of these will be the following:

(α) the axiomatics
(β) the core
(γ) the mathematical part
(δ) the domain of intended application
(ϵ) the empirical description

Components α through δ will be treated in this section, component ϵ in the next chapter.

(α) As a possible axiom system in the extension of set theory, we choose a version of so-called *species of structure* in the sense of Bourbaki (1968, Chap. 4) which is adapted to our set-theoretic framework. The items about which such a system of axioms speaks, are then finite systems

$$\langle X_1 \ldots X_m; A_1 \ldots A_l; s_1 \ldots s_n \rangle \tag{2.39}$$

of objects from the universe of sets of the previous section. The X_μ are physical, the A_λ mathematical, and the s_ν could be physical, mathematical or mixed, depending on their construction (see below). The particular natures of the s_ν, therefore, need not be specially specified since they are self-evident. But it is necessary that

$$ph(X_\mu), ma(A_\lambda) \tag{2.40}$$

for all μ, λ whereby the A_λ, in addition, have to be well-defined mathematical sets, i.e., constants of the theory (see β).

The so-called *principal base sets* X_μ are those sets whose elements are no longer 'explained' within the respective theory of (2.40), but rather are to be determined from the outside. That they are, in addition, physical according to (2.40) does not mean that for that reason they *actually* have to exist. It suffices that they are *really possible*. In the description of an actually present physical system, doubtless, some of the terms used have a real significance, e.g., *the* temperature, *the* energy, etc., that the system currently has. Likewise, unquestionably, there also exist in such a description always (at least implicitly) terms that suggest only real possibilities just not realized in the present case—the other temperatures, other energies, etc. We must accept this fact as long as our knowledge of reality is *conceptual*. The concept of a horse, as applied to only one horse ('this is a horse"), also includes indefinitely many states of horses, and certainly also those that are never realized.

In our example theory of Sect. 2.1, space and time (as point sets) are principal base sets, as well as the sets of possible forces. We indicated there that, strictly speaking, they would include the sets of possible physical values of spatial and temporal distances, masses, etc., but in order to simplify we have already replaced them by the set of positive real numbers and their possible numerical values. Here, we meet the so-called *auxiliary base sets* A_λ. We see these, as we have said, as well-defined mathematical sets in the sense of the previous section, so as $\mathbb{R}$ or the set $\mathbb{C}$ of complex numbers, $\mathbb{N}$ of natural numbers, the orthogonal group (as a matrix group), etc. Mathematical quantities of this type, thus, explicitly form a possible component of a physical theory, even of a physical system (2.39). Finally, we have the *typified sets* s_ν. They are elements of sets, so-called *ladder sets*, which are certain constructions of the principal and auxiliary base sets 'explained' in every case *within* the theory. Permitted constructions are the repeated applications of the operation

$$Power\ set\ of\ a\ Cartesian\ product$$

on the base sets and each set already produced. With σ_ν as such a rule, a *ladder term*, we obtain the statements

$$s_\nu \in \sigma_\nu(X; A) \tag{2.41}$$

which typify the s_ν (and hence their elements) via the ladder sets $\sigma_\nu(X; A)$. (Here, as below, we abbreviate the names for the systems of sets X_μ, A_λ, etc., by X, A, etc., if no misunderstanding is possible.) In the example theory of Sect. 2.1, the spatial metric d (with physical distances), the trajectory described by real functions $\mathbf{x}$, and a coordinate system ϕ would be typical cases of purely physical, purely mathematical, or mixed quantities s_ν with the typifications

$$d \in \mathrm{Pow}(X^2 \times L), \mathbf{x} \in \mathrm{Pow}(\mathbb{R} \times \mathbb{R}^3), \phi \in \mathrm{Pow}(X^3 \times \mathbb{R}^3) \tag{2.42}$$

(X the point space, L the set of possible distances). It is, therefore, known for the s of what elements they are composed, in so far as this is known through external determination of the X: One easily sees that the Cartesian product of two physical (mathematical) sets is once again physical (mathematical) and the same applies to the power set of a set. This results in the nature of a typified set in general.

Equations (2.40) and (2.41) (respectively for all indices) form the first two (propositional) constituents of a species of structure in our sense. Added to this are the actual axioms

$$\alpha(X; A; s)\ , \tag{2.43}$$

which make a largely arbitrary statement about the physical system (2.39)—largely, but not entirely arbitrary. As a matter of course, it can be the case that we want to have α consistent, and that would be achieved, say, through

$$\vdash \exists X, s : \alpha(X; A; s) \,. \tag{2.44}$$

The second requirement is an important invariance requirement, which the typification (2.41) itself fulfills. It regards the *canonical invariance* of α with respect to isomorphisms of structures.

If we proceed through bijections f from the sets X to the sets X', and if s is a set typified through X and A, then s corresponds uniquely to a set s' with *the same* typification through X' and A (which remain fixed), respectively, that s has through X and A, respectively. This makes itself readily clear through recursion over the formation of typifications. Denoting as *isomorphic* the structures $(X; A; s)$ and $(X'; A; s')$ just formed, one thus has that (with A suppressed)

$$\mathrm{isom}(X, s; X', s') \rightarrow [s_\nu \in \sigma_\nu(X) \leftrightarrow s'_\nu \in \sigma_\nu(X')] \tag{2.45}$$

is valid. The typifications are, therefore, statements associated with each structure $\langle X; A; s\rangle$ that are valid or invalid as well for each structure $\langle X'; A; s'\rangle$ isomorphic to it. In this sense, we learn nothing about the elements of the sets X by typifying: *salva veritate*, we can replace them by any others which are isomorphic to them.

This invariance does not hold absolutely for all conceivable statements α. We can indeed say about two principal base sets that they have no element in common. Or we can say of a principal base set that it contains the set of real numbers (as a set A). Such statements are immediately eliminated by isomorphisms. We must, therefore, demand of the actual specific axioms what is true for the typification by itself. And we want this, indeed: So,

$$\mathrm{isom}(X, s; X', s') \rightarrow [\alpha(X, s) \leftrightarrow \alpha(X', s')] \tag{2.46}$$

becomes the other condition we impose on a species of structure. In all this, it should be noted that the statements $ph(X_\mu)$ in (2.40) are not invariant under isomorphisms. So we have (2.45) for the typification and require (2.46) for α only *without* the requirement $ph(X_\mu)$ in (2.40). This requirement has very pleasant consequences with respect to the more specific and well-known invariance requirements of physics (cf. Scheibe 1994a).

To the axioms α permitted by the very weak conditions (2.44–2.46) belong all set-theoretic translations of type-logical statements, where the typifications (2.41) remain as residue of the logical types of the original predicates. The bound variables in the original language that, likewise, have a type would also be constrained on the corresponding ladder sets in such translations. Under these circumstances, the decision regarding the validity of α would depend only on the knowledge of the structure (2.39) and not as well on the universe $\mathcal{L}$ in which it is embedded. The bound variables would, in these circumstances, carry a label as physical, mathematical, or mixed. Such cases have the pleasant aspect that, for a physical theory, we do not take as the subject of the theory the entire physical (!) set universe, but only the ladder sets over *one* basis $\langle X; A \rangle$. Nonetheless, we will discuss these features occasionally in the following but do not require them to be fulfilled generally. We speak of species of structures only under the conditions (2.41, 2.43) and (2.44–2.46), summarizing (2.41, 2.43) conjunctively in a formula $\sum(X; A; s)$ to which in the applications (2.40) adds.

(β) Of the parameters X, A, and s that enter in the axiom system Σ some are considered *constants* and the others *variables*. In our standard example, the parameters for the description of space and time, as well as the force field, are constants, and we have even more physical constants in the theory if we also leave the sets of values for spatial and temporal distances, as well as for the masses, as distinct principal base sets and do not replace them from the outset by numerical scale sets. In contrast, the parameters m for the mass of the particle and $\mathbf{x}$ for its motion are variables, of course. While we moved to a new theory with a change of the force field, and because this new field is not an object but rather is part of the theory, m and $\mathbf{x}$ are variables *within* the theory. Just this is what makes the theory general. Although the demarcation between constants and variables of a theory is to a certain extent arbitrary, it is important to take this distinction explicitly into the notion of theory and to set it firmly down in each case, for a clear understanding and for avoiding unphysical cases.

In general, we will proceed in this matter in such a way that all basic terms are constants, with variables occurring only under the typified terms. The reason for doing so is, above all, that otherwise the axioms would (and had to) admit of too large classes of structures as physically possible systems, in particular, with regard to the principal base terms which have to be interpreted by physical entities. In certain cases, it is simply that we want to take into account only *one* structure (*including* the typified sets). This is certainly the case for space and time in Newtonian physics, if they are taken in each case as a *whole* (not just locally). In other cases, such as the solutions of an equation of motion, we want the multiplicity that, here, exactly constitutes the generality of the theory. But it seems that we can satisfy this claim of generality already by permitting variables only among the typified terms.

It is true that some important constructs of theoretical physics are called theories, even though they have variable principal base sets. This is certainly the case for classical Hamiltonian mechanics and quantum mechanics and, in a restricted sense, also for general relativity. But what we have here is, actually, nothing more than a particular *formalism*. Classical phase space and quantum mechanical state space, limited only by the conventional requirements, stand for a vast set of possibilities, and the associated formalism, in this generality, can simply not be regarded as physically interpreted. Even if one leaves the Hamiltonian function or the Hamiltonian operator variable in the respective theory, the said spaces must in content be defined in a certain way if we still want to speak of a *physical* theory. All options remain open, but for their utilization in *one* theory it is necessary to state which possibility is being dealt with.

We come, thereby, to a second piece of information about a physical theory which is added to its species of structure Σ (and thus its axioms): We distinguish as the *core* of a theory a structure $\langle X^\circ; A^\circ; s^\circ\rangle$ which is a common fragment of all models of Σ. We further distinguish a species of structure Σ° as the strongest statement in Σ over $\langle X^\circ; A^\circ; s^\circ\rangle$. Generality is not significantly limited if we do so in the form

$$\Sigma(\mathrm{X}; \mathrm{A}; \mathrm{s}, \mathrm{s}_1) \equiv \Sigma^\circ(X; A; s) \wedge s_1 \in \sigma_1(X; A) \wedge \alpha_1(X; A; s, s_1) \,. \tag{2.47}$$

For a structure $\langle X; A; s\rangle$ that satisfies Σ to respect *also* the core of the theory, the following restriction

$$X = X^\circ; A = A^\circ; s = \langle s^\circ, s_1\rangle \,, \tag{2.48}$$

as well as the empirical claim

$$\Sigma^\circ(X^\circ; A^\circ; s^\circ) \,, \tag{2.49}$$

must hold. The X°, A°, and s° are then constants, and only s_1 remains variable. In our example theory, the core consists of space, time, the 'force space', the set of mass values, and the force field, together with the assumptions already to be made *about them*. The equation of motion, which in addition includes m and $\mathbf{x}$ as variables, does not belong essentially to it.

(γ) Next, we turn to the auxiliary base sets $A_1, \ldots, A_l$ of the species of structure of a physical theory. They constitute mainly the *mathematical part* of the theory. We had already at the beginning, under (α), determined that the sets A, as defined in the previous section, must be mathematical in nature and treated them as constants in (β). This does not exclude their all belonging to the core of the theory or the occurrence of typified sets s that are considered variable and are mathematical in nature. Either way, the question now arises of whether the occurrence of mathematical entities in a physical theory belongs to the essence of physics such that it cannot do without this ingredient, and if so, wherein this indispensability is established and sustained (see Scheibe 1994a). Since Galileo's time, physicists have persistently expressed their astonishment that "the book of nature is written in the language of mathematics". Below, this astonishment will be neither explained nor made to disappear. For our purposes, it is sufficient to find out *about what we have to be amazed*. The matter is taken up here primarily because the occurrence of mathematical entities alongside physical entities in a theory complicates the business of reduction, which we shall sometimes overlook somewhat light-heartedly. So this circumstance should be pointed out to the reader right at the beginning, in order to enable him eventually to take improvements into account himself. (See below for the concept of axiomatic basis of a physical theory in Ludwig 1990, Sect. 7.)

Mathematical entities can already enter into a species of structure in the construction of the individual structure of a given species and, of course, in the axiom of the species of structure. But they may be missing in the structures and appear only in the axiom and, eventually, both may occur. Relevant cases of the first type in physics are, for example, defined by sets of local-coordinate manifolds, as well as real and complex vector spaces. Recently, attempts have been made in differential geometry to develop coordinate-free manifolds (e.g., Misner, Thorne, and Wheeler 1973, Emch 1984, Part II), and this certainly emerged from the need to work with physically interpretable terms as much as possible. The real numbers are *essential* to the notion of a (local) coordinate system since it makes no sense to ask by what physical entities they would ultimately be replaced. Rather, one *wants* in this case to give the description of a possible physical event by a quadruple of real numbers—one wants this to the extent that numbers in physical statements can be tolerated at all. But the elimination of coordinate systems has so far not been fully successful. A related case is that of the real or complex vector spaces, especially since tangential vector spaces and higher tensor spaces are included in differential geometry. In such cases, the numbers that occur are a legacy of the coordinate systems of the underlying manifold, and it is through scalar multiplication that the numbers are insolubly bound to the vectors.

The case in which mathematical entities occur only in the axiom of a species of structure, we illustrate first by a geometric example. That there are fragments of physical theories that *can* be kept entirely free of mathematical entities is—perhaps somewhat paradoxically—particularly evident in *Euclidean geometry (EG)*. The direct appeal to geometry here might be surprising, because it is that which has been regarded as a genuine piece of mathematics besides arithmetic from ancient times. But, on this point, we want to take the position which among physicists—though not philosophers of science—since Einstein is generally agreed, namely, that the geometry used in physics is just as empirical as the rest of physics. For example, the congruence between pairs of points and the between relation for points in space can, in principle, be introduced physical-operationally. The geometric statements expressing characteristics of these relations are then subject to empirical testing. In this way, geometrical propositions suffer the same fate as assertions about how a charged particle moves in an electromagnetic field: They can be empirically confirmed or refuted. And this situation calls for just treatment today, even though it has for centuries been considered differently. What in mathematics still remains in this connection is the determination of logical (and therefore truly non-empirical) relationships between propositions of geometry: that some of them follow from others. Einstein (1921) had viewed the matter in the same way. But, unfortunately, this is still not the whole story.

It is also necessary to clarify the question of what sorts of geometrical statements are allowed in the theory. And there are two very different possibilities. These have already been mentioned in Sect. 2.1 in relation to distance geometry. We can go the path of analytic geometry and, thereby, permit geometric assertions of type (2.8). In the case of a geometry with congruency and a betweenness relation, we accordingly have the requirement of existence of a coordinate system ϕ for the (whole) space, so that (with x_i^ϕ as coordinates of the points $\mathbf{x}$ in ϕ, etc.)

$$\langle\langle \mathbf{xy}\rangle\langle \mathbf{uv}\rangle\rangle \in \mathrm{cgr} \leftrightarrow \sum_i (x_i^\phi - y_i^\phi)^2 = \sum_i (u_i^\phi - v_i^\phi)^2 \tag{2.50}$$

holds for congruence and

$$\langle \mathbf{xyz} \rangle \in \mathrm{btw} \leftrightarrow \sqrt{\sum_i (x_i^\phi - y_i^\phi)^2} + \sqrt{\sum_i (y_i^\phi - z_i^\phi)^2} = \sqrt{\sum_i (x_i^\phi - z_i^\phi)^2} \quad (2.51)$$

holds for the betweenness relation. While these physical entities occur on the left side of this equivalence, the right side is, instead of points (left), about triples of numbers and, instead of geometric relations (left), it is about certain arithmetic relations. And it is obvious that we *want* to bring mathematical entities into play with these statements: We want to say that the geometric relations between points in space (left) behave as very specific arithmetic relations between number triples (right). It is essential that at left physical and at right mathematical relationships are meant because the spirit of the enterprise is to try to describe physical relationships by mathematical ones. But, unlike the first kind described above, this time the coordinate systems do not enter into the construction of the respective geometrical structure, which is only given by cgr and btw.

On the other hand, we are not obliged in this case to use the analytical method. Rather, it is actually more natural in *synthetic geometry* to formulate geometrical propositions *directly* without the intervention of non-geometrical entities and relationships, such as transitivity

$$\langle \mathbf{xyz} \rangle \in \mathrm{btw} \wedge \langle \mathbf{yuz} \rangle \in \mathrm{btw} \rightarrow \langle \mathbf{xyu} \rangle \in \mathrm{btw} \quad (2.52)$$

for the betweenness relation, or the segment construction

$$\exists \mathbf{z}.\langle \mathbf{xyz} \rangle \in \mathrm{btw} \wedge \langle \langle \mathbf{yz} \rangle \langle \mathbf{uv} \rangle \rangle \in \mathrm{cgr}., \quad (2.53)$$

or something else similar. Here, in addition to the geometric relations, only logical tools come into play, without which we could not work in any case.

In this way, one could try to get a full axiomatization of EG *either* through statements of type (2.50–2.51) *or* through those of the type (2.52–2.53). Indeed, in the former case, (2.50–2.51) was already sufficient while, in the latter case, a lot more must be done than is accomplished by (2.52–2.53). But it *is* also done with the result, in particular, that the two systems of axioms (within set theory) are *equivalent* (Tarski 1959). From the point of view of the analytical solution to the problem, this means that the use of mathematical entities and relationships *alongside* the physical is dispensable. However, it also means, conversely, that one can introduce pure mathematical quantities in a synthetic solution without changing the physical content. In fact, this is how it has been historically, and the analytical method has been welcomed, above all, because it has simplified many things. At least, we had to wonder (in the above question) about this, whereas independently of such additions the formulation of EG as a physical theory actually requires no mathematics in the narrow sense.

Can we generalize this result? We hardly expect that each species of structure is equivalent to one without auxiliary base sets, as already shown by simple examples with physically relevant structure. We want to embed the *van der Waals* law

$$(p + a/v^2) \cdot (v - b) = R \cdot T \quad (p, v, T \neq 0) \tag{2.54}$$

for a real gas in the following small theory: We include in the core of the theory the value sets X°, Y°, and Z° for pressure, volume and temperature of a gas, and also the set $\mathbb{R}$ of the real numbers, the rational operations on $\mathbb{R}$ and, finally, the gas constant R, as well as the van der Waals constants a and b, which we think of as already given by their numerical values. This leads to the typifications

$$R \in \mathbb{R}, \quad a \in \mathbb{R} \ldots \tag{2.55}$$

Only the parameters $\mathbf{p}$, $\mathbf{v}$, and $\mathbf{T}$ for pressure, volume and temperature, respectively, are variable in this theory, indicating the state of the gas and having the typifications

$$\mathbf{p} \in X^\circ, \mathbf{v} \in Y^\circ, \mathbf{T} \in Z^\circ . \tag{2.56}$$

The crucial axiom of the theory, then, is that it provides the scalars ϕ, χ and ψ for X°, Y°, and Z°, respectively, in a way such that the numerical values of $\mathbf{p}$, $\mathbf{v}$ and $\mathbf{T}$ satisfy the Eq. (2.54).

One sees that something similar happens here as happens in the geometric example, only this time in the variable part of the theory: Once the values of pressure, volume and temperature, which are according to their nature completely different, are replaced by numerical values and are, in this way, made uniform, taking the constants R, a and b into account all 3-place relations between real numbers may describe the relationship of the three variables. Equation (2.54) is one of them that works even with the rational operations on numbers. Again, the advantage of the *richer* mathematical structure over the physical is utilized. The difference from (2.50–2.51) is that the relationship expressed in (2.54) is, for its part, not a structural element of the gas considered, about which statements were still to be made, but already constitutes the law governing the behavior of pressure, volume, and temperature. The main point is that, this time, it is impossible to substitute the numerical relationship (2.54) by axioms that (in addition to logical auxiliaries) only employ physical concepts. Which concepts should these be, then? Something like this would only be conceivable in the context of a physical *explanation* of the van der Waals equation, for example, by the kinetic theory of gases, which would illuminate the conceptual relationships between the relevant variables. But that would mean posing the same question for a *different* theory. Here, however, it was first about the purely phenomenological theory with the basic law (2.54). And there is nothing to be done for it in the matter in question.

So we have something to wonder about here: A law of physics as (2.54) cannot be expressed by using physical concepts alone, but certainly must draw on the assistance of mathematical entities such as real numbers and elementary operations. However, it must not be forgotten about this wonder that, as emphasized in the previous section, we *want* to establish the relationship to mathematical entities thoroughly in specific instances—unless we are uncompromising nominalists (cf. Field 1980).

The specification of the number of an empirically given quantity was mentioned was an elementary example. But the specification of the numerical value of a quantity belongs here, as in the last example, even if a certain arbitrariness already enters here, as in the choice of the unit. When, then, on the way to formulating laws such as (2.54), more and more mathematical stuff is coming in, one begins to look again for physical concepts and is surprised when they no longer are at hand.

If we now proceed to a generalization of the two examples, it should not be concluded from this that *every* physical species of structure comes to be specified in this way. We want to present here only one typical case of the application of mathematics in *its* full generality. The basic idea is to obtain information about *unknown* physical structures of a, hitherto, not directly axiomatized type by recognizing them as isomorphic to some mathematical structures, which are generally deduced in turn from yet other mathematical structures of a *known*, directly axiomatized type. Thus, for the example the EG, the Euclidean space ⟨S; cgr, btw⟩ is initially isomorphic to the number space, according to (2.50–2.51), with the base set $\mathbb{R}^3$ and the typified quantities in (2.50–2.51) at right. This 3-dimensional Euclidean number space is, then, deducible from the usual structure of the real numbers, the theory of which is known (see, e.g., Tarski 1956, Ch. X). This arrangement eventually results in conclusions that can be expressed *solely* by means of the terms cgr and btw, via (2.50–2.51). These, then, are the desired geometric propositions (2.52–2.53).

The concept of *deduction* of structures and species of structure is seen, in closer analysis of the general situation, to establish itself as a key notion. Since we will need this concept later in a different context (especially in Sect. 4.4), we want now to consider it in itself. The point is that $\langle X; s\rangle$ defines a new structure $\langle Y; t\rangle$ on the basis of a species of structure $\Sigma'(X; s)$, and its own separate species of structure $\Sigma\langle Y; t\rangle$ derives from $\Sigma'\langle X; s\rangle$. Thus, we have the deduction

$$\Sigma'(X; s) \wedge Y = P(X; s) \wedge t = q(X; s) \vdash \Sigma(Y; t)\,, \tag{2.57}$$

wherein P and q are defining terms, so-called *internal terms*, of Σ'. As such, they satisfy

$$\begin{aligned} \Sigma'(X; s) &\vdash P(X; s) \in \rho(X) \\ \Sigma'(X; s) &\vdash P(X; s) \in \tau(X) \end{aligned} \tag{2.58}$$

with certain (systems of) ladder terms ρ and τ, and invariances

$$\begin{aligned} \Sigma'(X; s) &\vdash P^*(X; s) = P(X^*; s^*) \\ \Sigma'(X; s) &\vdash \ q^*(X; s) = \ q(X^*; s^*) \end{aligned} \tag{2.59}$$

enabled by (2.58). In this way, we move from the X to the isomorphic X^*, and the s^*, P^* and q^* are the corresponding transitions to the ladder sets. In order for Σ to become a species of structure with the principal base terms Y and the typified terms t, the typification of q must be secured through P, so that

$$\Sigma'(X; s) \vdash q(X; s) \in \theta(P(X; s)) \,, \tag{2.60}$$

with ladder term θ. Summarizing, with (2.57) as a map which relates to each structure $\langle X; s\rangle$ of species Σ' a structure $\langle Y; t\rangle$ of type Σ, so (2.59) states that, by this map, isomorphic structures go over again to isomorphic structures. We could, for example, move from a group $\langle X; s\rangle$ and assign to it the lattice $\langle Y; t\rangle$ of its subgroups. Then isomorphic groups go over to isomorphic lattices. The single lattice is deduced here from an originally given group, and the lattice property from the group property (together with the mapping rule).

In (2.57), we have drawn conclusions Σ from Σ', but how *strong* Σ is in relation to Σ' remains completely open. Σ is as strong as possible if, together with (2.59), the 'inversion'

$$\Sigma(Y; t) \vdash \exists Xs.\Sigma'(X; s) \wedge Y \approx P(X; s) \wedge t \approx q(X; s). \tag{2.61}$$

holds (with $\approx$ being 'isomorphic to'). One sees from this that, for the given Σ', P, and q, in a certain sense, one can not run into the difficulty of knowing no inferences in the sense of (2.57). For the right side of (2.61) *always* follows and is even the strongest conclusion because (2.57) always holds in the form

$$\begin{aligned} &\Sigma'(X; s) \wedge Y = P(X; s) \wedge t = q(X; s) \\ &\quad \vdash \exists Xs.\Sigma'(X; s) \wedge Y \approx P(X; s) \wedge t \approx q(X; s). \end{aligned} \tag{2.62}$$

However, the conclusion we obtain here takes an unusual form that one will be reluctant to accept when it comes to a proper 'understanding' of the statements made. In our example, one would like the statement that $\langle Y; t\rangle$ is isomorphic to a subgroup lattice preferably to have a form in which it is expressed by the direct use of the lattice operations. And, yet, the form in question is at least tentatively accepted in physics.

Exactly that happens, that is, when one uses analytic geometry, as in the small gas theory (2.54–2.56) and in countless other theories of physics, such as our example theory of Sect. 2.1. Taking into account the differentiation of a core in the sense of (2.47–2.49), we have the following general method for recourse to mathematics in a physical theory. We start from a mathematical species of structure that applies to mathematical structures A°, a° (constant), and a (variable):

$$\Sigma'(A^\circ; a^\circ, a) \,. \tag{2.63}$$

In addition, we employ mathematical terms P°, q°, and q from which we can, under assumption $\Sigma'(A^\circ; a^\circ, a)$, deduce the mathematical structures

$$\langle P^\circ(A^\circ; a^\circ), q^\circ(A^\circ, a^\circ), q(A^\circ; a^\circ, a)\rangle \,. \tag{2.64}$$

Those factors are finally applied to make the statement

$$\exists a.\iota \cdot Y^\circ = P^\circ(A^\circ; a^\circ) \wedge \iota \cdot t^\circ = q^\circ(A^\circ; a^\circ) \\ \wedge \iota \cdot t = q(A^\circ; a^\circ, a) \wedge \Sigma'(A^\circ; a^\circ, a). \tag{2.65}$$

about physical structures $\langle Y^\circ; t^\circ, t\rangle$. Here ι is an isomorphism (already continued on the ladder set), and we just say that the physical structure $\langle Y^\circ; t^\circ, t\rangle$ is isomorphic to the mathematical structure (2.64) whose arguments $A^\circ; a^\circ, a$ in turn satisfy (2.63). Thus, we take the sense of (2.62) to be the strongest statement that is deducible from (2.63–2.64). Regarding the variability of t and a, it is to be noted that it is *one* isomorphism, that accomplishes the transition from physics to mathematics for all possible t and a. We finally want to mention two examples in the light of generalization obtained. In the first, quantum mechanics, the variable part in (2.63–2.64) is trivial, and the emphasis is on the core. In the second, electrodynamic example, it is vice versa.

The structure commonly associated by physicists with quantum mechanics is the Hilbert space of square-integrable complex spatial functions. The presentations more focused on fundamental issues, however, often express the view that two proportional Hilbert vectors are physically indistinguishable (see many of the contributions to Hooker 1975, 1979). It is, therefore, not the Hilbert space itself that can take over the role of a physical structure here (up to isomorphism). Rather, it acts as auxiliary mathematical structure in the background and allows the deduction of certain other structures to which, according to (2.65), the physically relevant structures are then isomorphic (see the particularly clear presentation in Mackey 1963, 71, Axiom VII). In the constant part (core), to them belong as structural elements, for example, the set G of self-adjoint linear operators for mathematical representation of the physical observables, the set Z of positive-trace bounded operators to represent the statistical states, the function

$$Tr : Z \times G \to \mathbb{R}, \quad E = Tr(w \cdot A) \tag{2.66}$$

(with Tr as a trace) to represent the expectation value function, etc. In the variable part of the axiomatics, we have (mathematically) the Hamiltonian H and the 'temporal' variation of the statistical operator W, which require the generalized Schrödinger equation

$$\hbar i \cdot dW/dt = H \cdot W - W \cdot H\ . \tag{2.67}$$

As in the gas example, pressure, volume, and temperature can be related by representing all three of them by real numbers (as their numeric values) so, here, the energy can be connected with the change of state by representing observables and states by operators on the same Hilbert space. In both cases, it is then simple algebraic operations which—in the second case together with a differential operation—allow the key equation to be formulated.

Indeed, while in quantum mechanics the constant terms Z, G, Tr, etc., but not the variables H and W, have to be derived, in classical electrodynamics this relation is reversed: There, one always finds next to the field strengths **E** and **H** the potentials ϕ and **A** on flat Minkowski spacetime (trivial terms P°, q° in (2.64)), but also the remark that two potentials

$$\mathbf{E} = -\text{grad}\phi - 1/c \cdot d\mathbf{A}/dt, \quad \mathbf{H} = \text{rot}\mathbf{A}\,, \tag{2.68}$$

which belong to the same field strengths, in the sense of the deduction (term q in (2.64)), are physically indistinguishable. And this is the motive in turn to banish, this time in the variables, the potentials from the role of sometimes convenient means of representation in a mathematical background theory, from which, via (2.68), only the physically relevant field strengths (up to isomorphism) were first obtained. The Maxwell equations for the field strengths with (2.68) then follow from the equations for the potentials. Here it is also important that, conversely, according to (2.68) every solution of Maxwell's equations can be represented by potentials.

(δ) Among the already distinct components of a theory, we now want to add the *domain of application*. Several problems are associated with this notion, which we can only survey. First, it is questionable whether the domain of a theory is one of its components. Because it changes over time and one wants the theory itself to be something that remains temporally unchanged—at least after a given time of consolidation, and even already during the time its domain of application, is being extended through testing. Also, there is *the* domain of application of a theory not in the sense, for example, that we would say of mechanics that it is applied in various fields. The largest problem connected with the question concept is rather the answer to the question of how a domain of application is to be *characterized.* Such characterizations must be *independent* of the specification of the relevant theory, as carried out according to the previous definitions (α) to (γ). For one thing, by the domain of application of a theory we mean by no means its *scope.* A theory is applied repeatedly even without its being known whether it is valid in this application. Also, one wants, for example, of classical mechanics to be able still to say that it has been applied to both atoms and ordinary matter, even if only *in part* successfully. Above all, the question of the language in which we should provide the domain of application belongs to the characterization. From a logical point of view, the physical systems of which one wants to assert the validity of the theory are gathered in the domain of application. But about which objects do we wish, for example, Kepler's laws to make assertions? If we wished to respond in the language of Kepler's laws, hardly more would came from these laws than stating of the objects that they satisfy Kepler's laws. To avoid this tautology, we must have recourse to a language that does not already anticipate the respective theory, but says something about how the respective class of objects is *given* to us. This can by be done by means of pointing, by enumeration, by specifying the scale, by imposing constraints (dilute gases, small speeds, etc.), and, in any case, by prescriptions as vague as are possible via the words body, atom, star, gas, crystal, etc.

These circumstances are to be considered if we now regarding a domain of application $\mathcal{R}$, selected in one way or the other, use the species of structure Σ provided in (β) to express the claim of validity of the theory not only, as before, for *one* system, but for all systems in $\mathcal{R}$, via

$$\forall \check{r}.\check{r} \in \mathcal{R} \to \Sigma(X^\circ; s^\circ, s^{\check{r}}) . \tag{2.69}$$

Thus, we understand (2.69) as an empirical claim, according to which for all systems $\check{r}$ belonging to $\mathbb{R}$, the associated structure $\langle X^\circ; s^\circ, s^{\check{r}}\rangle$ (assumed by the theory) satisfies the statement Σ. The variability of these structures is limited by the fact that $\langle X^\circ; s^\circ\rangle$ was taken to be firmly bound to the theory. According to (2.47), (2.69) claims in addition that

$$\Sigma^\circ(X^\circ; s^\circ) \tag{2.70}$$

is empirically valid for the core of the theory. For the remaining structures $\langle \ldots; \ldots s^{\check{r}}\rangle$, by contrast, a different claim results each time.

2.4 Concepts of a Theory

Within a physical theory, in the sense developed so far, lie entities that we are inclined to classify as *concepts*. Those entities, of which we would say that they are "the concepts of the theory," are only one of several possible sorts of concept formation in connection with the theory. But when it comes to questions of reduction, it is exactly these concepts that are on trial. In particular, difficulties in the reduction of a theory T to another theory T' are often shifted on to the concepts of the respective theories. Then they say, for example, that the terms of T' are incommensurable with those of T or the terms of T cannot be gained from those of T', etc. So we must take care, in any case, as to the membership of concepts to a theory we intend, when we speak simply of *the* concepts of a theory. We will shortly touch upon the very different type of notion we have *of* a theory.

Here, this regards the concept of the *model* of a species of structure, that is, those structures of the set universe which are of their kind or which fulfill the axiom system in question. More precisely, it regards the pair consisting of the species of structure as an axiom system and the associated model class. The latter is the scope of the concept, the former its intension. In mathematics, we sometimes have names for both the species of structure and its models, such as Riemann geometry and Riemannian spaces, topology and topological spaces, etc. In other cases, we have names only for the models and, then, rename the theory (species of structure) after it, for example, the theory of differentiable manifolds, the theory of Banach spaces, of Lie groups, etc. Directly connected with these *absolute concepts*, as they could be called, there are certain less well-defined terms, which constrain the absolute concepts to other such concepts. One speaks of the concepts of the commutativity (a ring), simplicity (a group), compactness (a topological space), etc. This is obviously not about concepts that have independent structural classes as their scopes. Rather, a species of structure must already be given so that such terms can be meaningfully applied: If we already know what is a ring, what a group, a topological space, etc., then we can define those other terms and apply constraints with their help. These terms are, thus, *attributive*, in that they provide for already known object classes attributes which limit them.

Absolute concepts stand in opposition to *relative concepts*, to concepts of (relative to) a theory. At the beginning of this chapter, we pointed out that the geometrical concepts of the circle, straight line, tangents, etc., do not stand alone. They must be supplemented not only by the space *in* which they are circles, lines, tangents, etc., but also by the associated geometry, so Euclidean space. Even the relative terms have, next to their extension, an intension: Once a *specific* space is given, there are not only the sets of circles, lines, etc., in it, but for *any* space that satisfies some of Euclidean geometry, *each* of these scopes is uniquely determined. More precisely and more generally, a relative concept is the mapping rule that maps a set, typified through its principal base sets, to each model of a given structure species, as the scope of the concept in the model in such a way that it, furthermore, (canonically) maps an isomorphic scope to isomorphic models. So the concept of geodesic (in the sense of the shortest) line is the rule to map to every Riemannian space the set of its geodesic lines so that isomorphic spaces receive isomorphic sets of geodesic lines. Formally, and in general, this mapping is made through an internal term q of type τ which, as an internal term, has the two properties (2.58–2.59) of the previous section. The hereby required properties of q, thus, generally relate essentially to a given structure species Σ (referred to Σ' in those formulas), and precisely this expresses the relational character of the concept in question. Of course, we can also make $\langle\tau, q\rangle$ an absolute concept by adding Σ so as, for example, to speak of the notion of circle in a Euclidean space. However, within this construction, the relational character of $\langle\tau, q\rangle$ relative to Σ is also preserved. On the other hand, it should be noted that the core of a theory limits its concepts (in its relativity) in the previously intended sense that the mapping remains restricted by q to those models which have the relevant core as a fragment. It is just this limitation by which the domain of the concepts of a theory (and the theory itself) becomes manageable.

The simplest concepts of a theory $\Sigma(X; s)$ are its *fundamental concepts*, defined by

$$q(X; s) \equiv X_i \text{ or } s_k . \tag{2.71}$$

It should be noted that the s, although typified by the X, are also fundamental concepts. In the direct axiomatizations, which are not done via set theory, even only the (equivalent of the) s appear, while the X are represented by variable types. For the remaining concepts, there are two manners in which a concept $\langle\tau, q\rangle$ can be formed. In both cases, we have an expansion (cf. 4.5)

$$\Sigma'(X; s, t) \equiv \Sigma(X; s) \wedge t \in \tau'(X) \wedge \alpha'(X; s, t) \tag{2.72}$$

of Σ. Then, first

$$\exists! y.\forall t.\Sigma'(X; s, t) \leftrightarrow t \in y.. \tag{2.73}$$

always holds, that is, the set of t with $\Sigma'(X; s, t)$ (with given $X; s$) uniquely exists. This follows from the typification of t together with the ZF axiom of comprehension. On the basis of (2.73), we also have

$$q(X; s) \equiv \{t|\Sigma'(X; s, t)\} . \tag{2.74}$$

In this case,

$$q(X; s) \in \text{Pow}(\tau'(X)) , \tag{2.75}$$

and $q(X; s)$ always has elements within the ladder sets over X. This is not necessarily so in the other case, in which, for *specific* Σ',

$$\exists! u . \forall t . \Sigma'(X; s, u) \tag{2.76}$$

is provable. It is then the basis for the construction of

$$q(X; s) \equiv \text{the } u \text{ for which } \Sigma'(X; s, u) , \tag{2.77}$$

from which, in this case, it follows that

$$q(X; s) \in \tau'(X) . \tag{2.78}$$

If $\tau'(X) = X_i$ for a value i, it may happen here that $q(X; s)$ has no element in the accepted framework.

The method of recovery of concepts of a theory according to the method of (2.73–2.75) is the normal case with "qualitative concepts". Examples, in addition to the already noted types of figures in Euclidean space, such as circles, ellipses, cubes, etc., are terms of the theory of differentiable manifolds, such as the tangent vector, cotangent vectors, and the higher order and mixed tensors, such as the corresponding fields, e.g., tangent-vector fields. To method (2.76–2.78) first belong the *singular* concepts, for which $\tau'(X) = X_i$ and (due to $q(X; s) \in X_i$) $q(X; s)$ has the meaning of an individual, which has no more elements inside of the ladder sets on X. For example, the length a of the semi-major axis is put in relation to the orbital period of the planetary motion in Kepler's third law,

$$a^3/T^2 = \mu/4\pi^2 . \tag{2.79}$$

In Kepler's theory, a and T are defined terms that depend on the motion $\mathbf{x}$ of the planet: $a = a(\mathbf{x})$ and $T = T(\mathbf{x})$. These terms have values in the principal base set of spatial and temporal distances, and these values are uniquely determined on the basis of the theory: The planet moves periodically on an ellipse and $\mathbf{x}$, therefore, uniquely takes on the values $a(\mathbf{x})$ and $T(\mathbf{x})$. It formally concerns singular concepts, and it also intuitively it makes no sense to ask for elements of the respective lengths and times. The situation is similar with all quantities of a physical system in a theory such as, for example, many-particle mechanics the (instantaneous) center of gravity, the total momentum, energy, etc. As part of a sufficiently complete description of the system, such quantities have uniquely determined values that lie in a principal base set and therefore themselves have no relevant elements.

But there are now also cases (2.76–2.78) that are not singular. If we have, for example, a Euclidean space in which three non-collinear points are selected, these uniquely determine the circle on which the three points lie. Of course, it is not in this case about the concept of the circle, but it concerns the concept of three given *points* having the same distance from another point. The perimeter of the given triangle is just now *itself* the concept of these points. And this concept is not singular. Another example is the affine connection, which in a Riemannian space preserves the length of the tangent vectors under translation. This connection is uniquely determined by the metric and, for that very reason, is a concept of Riemannian geometry according to (2.76–2.78). Of course, it is as little the concept of affine connection in general as that perimeter of the given triangle was the concept of circle in general. The corresponding general concepts already belong to the theory of differentiable manifolds and Euclidean geometry, respectively, without a Riemannian metric or a triangle being chosen. On the other hand, the definite affine connection is not a singular concept. Its extension depends upon which sort of affine connection we have before us and in the case of the version of Koszul (for example) would be a set of triples of vector fields, one of which is the derivative of the second with respect to the third (cf. Westenholz 1978, Sect. 11.2). Bearing in mind the general rule, according to which we denominate a concept according to the *objects* that fall under it, one would have to say here that the singular such affine connection would be the concept of a (covariant) derivative.

In this section, we have often brought examples of concepts that could be construed as *mathematical* concepts according to the ordinary use of language, such as the notion of a tangent vector as a concept of the theory of differentiable manifolds. But this ordinary usage does not always meet our specifications for the difference between physical, mathematical, and mixed objects of our 'universe of discourse' nor respective specifications concerning physical and mathematical statements (cf. Sects. 2.2 and 2.3). According to standard views, the concept just mentioned is mathematical at most due to its generality but, above all, because it is part of an education in mathematics. According to our views, it (just due to generality) is rather a mixed concept which would lead only to a mathematical concept by limiting it to manifolds of mathematical objects.

If we look for a term $q(X; s)$ restricted to typified sets s, which are non-mathematical, all the concepts coming from (2.71) through (2.78) are physical. Then, given the other conditions, the objects $q(X; s)$ are also at least mixed, and this is what we have to be satisfied with in a setup that allows, in principle, that there exist physical systems consisting of physical *and* mathematical elements. The rest is taken care of by our provision that a physical theory restricts its species of structure always to a core with physical main base sets. It follows that there is no physical theory of differentiable manifolds or the like. Only when a certain manifold is (by idealization) fixed as a physical object, do we come so far in proximity to a physical theory. And then the general concept of the tangent vector is restricted accordingly, and becomes a physical concept.

References

Balzer, W., Moulines, C. U., and J. D. Sneed 1987. *An Architectonic for Science. The Structuralist Program*. Dordrecht: Reidel.

Bourbaki, N. 1968. *Elements of Mathematics. Theory of Sets*. Reading, MA: Addison-Wesley. French version:

Bourbaki, N. 1970. *Théorie des ensembles* 1 á 4. Paris: Hermann 1970. Repr.: Paris: Masson 1990.

Carnap, R. ^{2}1959. *Introduction to Semantics*. Cambridge, MA: Harvard University Press.

Da Costa, N. C. A., and F. A. Dori. 1992. Suppes Predicates for Classical Physics. In: J. Echeverria et al. (Eds.), *The Space of Mathematics*. Berlin: De Gruyter, 168–191.

Dummett, M. 1978. *Truth and Other Enigmas*. London: Duckworth.

Einstein, A. 1921. Geometrie und Erfahrung. Repr. in: Einstein, A.: *Mein Weltbild*. Ed. by C. Seelig. Ullstein Materialien: Frankfurt 1989, 119–127. Extended version: *Geometrie und Erfahrung. Erweiterte Fassung des Festvortrages Gehalten an der Preussischen Akademie der Wissenschaften zu Berlin am 27. January 1921*. Berlin: Julius Springer 1921.

Emch, G. G. 1984. *Mathematical and Conceptual Foundations of 20th-Century Physics*. Amsterdam: North-Holland.

Field, H. 1980. *Science without Numbers. A Defence of Nominalism*. Princeton, NJ: Princeton University Press.

Fraenkel, A. A., Y. Bar-Hillel, and A. Levy. ^{2}1973. *Foundations of Set Theory*. Amsterdam: North-Holland.

Hooker, C. A. (ed.). 1975. *The Logico-Algebraic Approach to Quantum Mechanics. Vol. I: Historical Evolution*. Dordrecht: Reidel.

Hooker, C. A. (ed.). 1979.*The Logico-Algebraic Approach to Quantum Mechanics. Vol. II: Contemporary Consolidation*. Dordrecht: Reidel.

Ludwig, G. 1985. *An Axiomatic Basis for Quantum Mechanics, vol. 1: Derivation of Hilbert Space Structure*. Berlin: Springer.

Ludwig, G. 1990. *Die Grundstrukturen einer physikalischen Theorie*. Berlin: Springer.

Mackey, G. W. 1963. *Mathematical Foundations of Quantum Mechanics*. New York: Benjamin.

Misner, C. W., K. S. Thorne, and J. A. Wheeler 1973. *Gravitation*. San Francisco: Freeman.

Scheibe, E. 1979. On the Structure of Physical Theories. In: I. Niiniluoto und R. Tuomela (Eds.), *The Logic and Epistemology of Scientific Change*. Amsterdam: North Holland, 205–24. Repr. in: Scheibe 2001, 160–174. German translation: Über die Struktur physikalischer Theorien. In: W. Balzer und M. Heidelberger (Eds.), *Zur Logik empirischer Theorien*. Berlin: de Gruyter 1983, 169–188.

Scheibe, E. 1982a. A Comparison of Two Recent Views on Theories. *Metamedicine* 3, 233–253. Repr. in: Scheibe 2001, 175–194.

Scheibe, E. 1986a. Mathematics and Physical Axiomatization. J. Vuillemin (Ed.), *Mérites et Limites des Méthodes Logiques en Philosophie*. Coll. Int. Fondation Singer-Polignac, Paris 1984. Paris: Vrin, 251–277. Repr. in: Scheibe 2001, 535–552.

Scheibe, E. 1994b. A Most General Principle of Invariance. In: E. Rudolph und I. -O. Stamatescu (Eds.), *Philosophy, Mathematics and Modern Physics. A Dialogue*. Berlin: Springer, 213–225. Repr. in: Scheibe 2001, 501–512.

Scheibe, E. 1994c. On the Mathematical Overdetermination of Physics. In: E. Rudolph und I. -O. Stamatescu (Eds.), *Philosophy, Mathematics and Modern Physics*. Berlin: Springer 1994, 186–

199. Revised version: The Mathematical Overdetermination of Physics. In: E. Agazzi and G. Darvas (Eds.), *Philosophy of Mathematics Today*. Dordrecht: Kluwer 1997, 269–285.
Scheibe, E. 2001. *Between Rationalism and Empiricism. Selected Papers in the Philosophy of Physics*. Ed. by B. Falkenburg. Berlin: Springer.
Shoenfield, J. R. 1967. *Mathematical Logic*. Reading, MA: Addison & Wesley.
Sneed, J. 1976. *Philosophical Problems in the Empirical Science of Science: A Formal Approach*. Erkenntnis 10, 115–146.
Suppe, F. (Ed.) 1974. *The Structure of Scientific Theories*. Urbana, Ill. University of Illinois Press.
Suppes, P. 1960. *Axiomatic Set Theory*. Princeton, NJ: Van Nostrand.
Tarski, A. [7]1956. *Introduction to Logic and to the Methodology of Deductive Sciences*. New York: Oxford University Press.
Tarski, A. 1959. What is Elementary Geometry? In: L. Henkin, P. Suppes, and A. Tarski (eds.), *The Axiomatic Method*. Amsterdam: North-Holland, 16–29.
Van Fraassen, B. 1980. *The Scientific Image*. Oxford: Clarendon.
Westenholz, G. V. 1978. *Differential Forms in Mathematical Physics*. Amsterdam: North-Holland.
Zermelo, E. 1908. Untersuchungen über die Grundlagen der Mengenlehre. *Mathematische Annalen* 65, 261–281.

Chapter 3
Corroboration and Empirical Progress

In this chapter, we get for the first time to our objective: reductions. However, we initially deal only with a very special kind of such thing, which we call the *empirical reduction*. When one confronts a physicist with two theories T and T', for which it holds that each phenomenon of T can be explained by T' and, what's more, that T' may also explain some phenomena inexplicable by T, perhaps he would already consider this relationship of T to T' to be a reduction of T by T'. At least this is what someone might think who considers the explanation of phenomena not only the most important, but even the *sole* task of a physical theory. In this view, it would consequently not be expected that anything *else* of physical significance of the theory T be inherited by T', and the relationship in question would appear to end up being the *only* type of reduction that exists at all. This view is not represented here. Our concern is precisely much more with drawing attention to very different types of reductions in accordance with the pronouncements of Sect. 1.3. Thus, even if there is between T and T' the indicated empirical reduction relation, it could still be *established* even further and in different ways. The empirical reduction could be *on the basis of*, for example, a generalization from T to T' or a refinement of T to T', or whatever types of reduction we encounter. It would then complete all these different types of reduction, and could serve as empirical justification for the other more specific ones. Such a completion would be more welcome anyway, insofar as the concept of empirical reduction would be a concept of *empirical progress* in the reverse direction: T' is an advance over T if T is empirically reduced to T' in the sense mentioned. If it could be accomplished to prove that every reduction is possible empirical progress, one would therefore have made useful the whole business of the reduction for the understanding of theory formation, and put it in a term that has shaped the development of modern physics like no other.

The outline just given of what empirical progress and empirical reduction are is now (among other things) still imprecise, insofar as we are in the following not concerned with the explanation of phenomena by a theory but with the more modest task of confirmation of a theory by observation and measurement. How close

E. Scheibe, *The Reduction of Physical Theories*, Fundamental Theories of Physics 207,
https://doi.org/10.1007/978-3-662-65000-4_3

together these two things lie depends on the respectively relied-upon explanation and confirmation concepts and, in at least *one* choice of these concepts, they move, in fact, very closely together: for the concept of D-N explanation and the concept of hypothetico-deductive confirmation. In both cases, it is above all that, from the relevant laws of the theory along with some elementary, contingent propositions, other such are deduced and their consistency with the measurements is determined. We want to make this close relationship the basis of our theory of corroboration (Sects. 3.1 and 3.2) which, in turn, will form the basis for the concept of empirical progress: T' would then be an empirical advance over T if any empirical success of T in terms of the hypothetico-deductive confirmation is also a success of T', and if there are successes of T' that are not successes of T (Sect. 3.3). We will, however, above all, have to ensure that beside the exact confirmation concept a notion of approximate confirmation is also available, which should take into account not only the fact that all measurements involve some inaccuracy, but also contributes to making the concepts of reduction and progress sufficiently liberal: A better, reducing theory must be able to include a real correction of the theory that has been overcome.

3.1 Hypothetico-Deductive Corroboration

The fundamental idea of the hypothetico-deductive corroboration of a physical theory is the possibility of developing, from its laws together with further premises, singular (or: contingent) propositions the validity of which can be empirically decided as directly as possible. If, in addition, there are good reasons for the validity of these additional premises, we thus arrive at a *test*—an empirical examination—of the theory in question: It turns out in the positive if the inferred statement is empirically valid and in the negative if not. For this procedure, it is crucial that the additional premises and the inferred statement are of a different kind than the laws of the theory: Only because they lie closer to our empirical grasp than these, is it possible to achieve the intended purpose of the empirical examination of the laws. We must, therefore, in particular, take care with new statements coming into play. In a quite specific case, they are known in physics as "initial and boundary conditions". (Compare the following with the orthodox treatment of the subject, for example, in Lenzen 1974, esp. Part I.)

Let us orient ourselves next with a few examples! In *Euclidean geometry* (EG), formulated as distance geometry, and the basis of our example physical theory of Sect. 2.1, on the one hand, we have theorems such as the (topological) metric theorems characterizing the distance d

$$\begin{aligned}
&d(x, y) \geq 0 \\
&d(x, y) = 0 \leftrightarrow x = y \\
&d(x, y) = d(y, x) \\
&d(x, y) + d(y, z) \geq d(x, z) ,
\end{aligned} \tag{3.1}$$

which are on the one hand quite simple but, on the other hand, pose difficulties for their empirical examination mainly due to their intended generality, i.e., that they should hold for all points $x, y, \ldots$ of space. Due to this complexity, the question of validity must be approached by way of the logical structure of (3.1). This could lead to success if, for our problem, we are better off with the singular statements of the form

$$d(x^\circ, y^\circ) = \alpha^\circ \text{ respectively } \neq \alpha^\circ;\ x^\circ = y^\circ \text{ respectively } \neq y^\circ \tag{3.2}$$

than with (3.1) and other theorems of EG. And we are: The empirical specification of geometry is, after all, done by specifying a procedure for measuring, for example, distances between points in space. Such procedures are (in principle) available to us *independently* of the validity of the theorems of EG and can, therefore, be used to test them. For the moment, we even want to assume that distance measurements can be carried out *exactly*, i.e., without measurement inaccuracies. Later, we will relax this assumption (cf. Sect. 3.2). But if we make it, then we will obtain statements of the form (3.2), which we may rightly assume to be valid if applicable. Here, the negated statements are not very informative, but it will turn out to be useful to include them.

The duality of lawful and contingent statements of EG given by (3.1), and (3.2) is carried on when we enter the more physical part of our example theory from Sect. 2.1. There, we had the law

$$md^2x_i/dt^2 = \alpha f_i(x) \tag{3.3}$$

for the motion $\mathbf{x(t)}$ of a particle and, again, already due to the generality of (3.3) with regard to time, an empirical test of this law as a whole is not possible. In addition, there is the difficulty that the force field $\mathbf{f(x)}$ is thought to be given in the whole space (and must be, see below), but, as such and strictly speaking, it is not an empirical datum. If we ignore this for the moment and take $\mathbf{f}$ as given, then we would have statements about the particle that correspond to (3.2), statements of the form

$$\begin{aligned}
&m = m^\circ \text{ respectively } \neq m^\circ ,\ x_i(t^\circ) = x_i^\circ \text{ respectively } \neq x_i^\circ \\
&\alpha = \alpha^\circ \text{ respectively } \neq \alpha^\circ ,\ dx_i/dt(t^\circ) = v_i^\circ \text{ respectively } \neq v_i^\circ
\end{aligned} \tag{3.4}$$

in which m°, α°, t°, $x_i(t^\circ)$, and v_i° are again (exactly) measured values. Again, we can imagine that the determination of such data, and thus the justification of statements (3.4) *can* (though need not) be done *independently* of the validity of law (3.3) (and a fortiori of the geometric laws) and may be used to test the law under this condition. As in the previous example, it should be noted that the description of a physical system by (3.2) or (3.4) significantly involves entities (the data) that do not appear in (3.1) or (3.3) and that, accordingly, new symbols must be introduced in order to achieve a contingent description. Unlike in the previous example, this time there are data involved with the velocities $\mathbf{v}^\circ$, which also contribute to the specification of the system but are not simply elements of the base sets.

Suppose now that, following the pattern of (3.2) and (3.4), we have *generally* a set of finitely many *observation statements*

$$A_1(x_1^\circ, \ldots, x_n^\circ), \ldots, A_m(x_1^\circ, \ldots, x_n^\circ)\,, \tag{3.5}$$

with *observational data* $x_1^\circ, \ldots, x_n^\circ$, for some physical theory. That we are concerned with observation statements and observational data is to say that we know experimental procedures to decide on the truth or falsity of these statements, and to provide the data *independently* of the validity of the theory in question, though not independently of *any* theory. More precisely, we think of the A_i as the result of a *test* of the theory in which m independent statements are decided and the A_i are, one by one, those of the B_i or $\neg B_i$ that turned out to be true. We first consider cases in which the single observation statements (3.5) at issue are *either* of the form

$$\langle \ldots x_k^\circ \ldots \rangle \in s_j \text{ respectively } \notin s_j\,,\; x_i^\circ \;=\; x_i^{\circ\prime} \text{ respectively } \neq x_i^{\circ\prime} \tag{3.6}$$

for elementary (!) terms s_j of the species of structure of the theory, which are at least linearly typified in Pow and the *elements* of which are n-tuples of possible empirical data, *or* of the form

$$s_j = x_j^\circ \text{ respectively } \neq x_j^\circ \tag{3.7}$$

if the elementary (!) s_j are *themselves* possible empirical data, which occurs, in particular, if they are elements of principal base sets and thus have no physically significant elements themselves. We then call the s_j in (3.7) and (3.6) of 1st an 2nd *order*, respectively. Later, it will become important that the s_j can also be non-elementary, defined terms (cf. Sect. 4.4). For the time being, the statements and data (3.5)–(3.7) may refer to *one* physical system, as it is initially the only object of a theory according to Sect. 2.3.

That we call the statements (3.5)–(3.7) 'observation statements' must not be understood too literally. It is well known from the history of recent empiricism that the attempt to give an absolute empirical basis to all empirically secured theorising as a foundation has led to apparently insuperable problems and has been abandoned for the time being. Accordingly, our observation statements are not supposed to derive from an empirical basis, but may well be arrived at with recourse to a theory. Once this is admitted, it no longer makes sense to distinguish between observation statements and theoretical statements in general (cf. Hempel 1965, Sect. 8; Carnap 1958). Rather, one must hope that one day a kind of hierarchy of all physical theories can be established in such a way that the observation statements of a theory T can only be made with the help of theories that are *prior* to T in this hierarchy. (Ludwig 1990,

Sect. 9.4 then speaks of "pre-theories"). Moreover, it must be shown in what sense this hierarchy has a basis of elements that no longer have any predecessors.

In addition to the problem of the basis, such an undertaking faces the problem of whether it is possible to avoid the fact that in the hierarchy some theories are necessarily their own predecessors, in that certain of their terms cannot be empirically determined in any other way than by reference to this very theory (cf. Sneed 1971). This would fail to meet our requirement that it must be possible to decide on the observation statements, which are to be used to test a theory, independently of this theory. We will also disregard this difficulty in the following. The weakening of the empirical content of such theories, should they exist, has been exaggerated (cf. Stegmüller 1976). It is clear that, e.g., within phenomenological thermodynamics, the gas constant R cannot be measured other than by invoking a gas equation, e.g., the equation for an ideal gas employing extremely dilute gases. Apart from the collection of sample gases for the determination of R, however, for any other gas the value of R can be regarded as externally determined, and the gas equation (with this value) can be tested again.

So, leaving aside these fundamental problems, our question now is: what do we know, if we have true observation statements (3.5)–(3.7), about the validity of the theory for the system in question—that is, in the above examples, about the validity of (3.1) and (3.3)? The answer will depend on the logical relationship between the observation statements and the axioms of the theory. Let us start with the worst case, that they *contradict* the theory. E.g., an empirical statement of the form

$$d(x^\circ, y^\circ) = \alpha^\circ, \quad d(y^\circ, x^\circ) \neq \alpha^\circ \tag{3.8}$$

would contradict the third theorem of (3.1) and, thus, EG. Some would say that in such a case the theory is falsified, and, according to an idea of Popper's, this is even the only thing we can definitively establish about a physical law (Popper 1935, passim). However, we do not want to be all that radical and, in these cases, speak only of the fact that the observation statements (3.5) constitute an *empirical failure* of the theory. This is to suggest that isolated empirical failures of a theory can be localised and generally leave large parts of the theory intact. A falsification of the parallel axiom is a falsification of EG, but in itself allows a theory very similar to EG to hold. Yet, even apart from the measurement inaccuracies to be considered in the next section, which make failures even worse, it must be said that already one precise empirical failure (if it existed) would seriously put strain on a theory and could not be tolerated as such in the long run.

The situation is less clear in the positive case, i.e., if the statements (3.5) are *compatible* with the theory. In this case, it may even happen that, no matter how much empirical material is accumulated according to (3.5), no *more* can ever be said than that this material is compatible with the theory. If we arbitrarily restrict the observation statements for EG to only two statements (3.2), we would have this situation, e.g., for the triangle inequality of (3.1)—simply because of the structure of this law: Less than three statements (3.2) do not yet challenge it, so to speak. To meet this challenge, one could add to the compatibility condition the further requirement that at least one variant of (3.5) contradicts the theory. Here, a variant is a set of contingent statements that emerges from the given ones by arbitrarily permuting the affirmations and negations. Our requirement, thus, boils down to the condition that a negative outcome must be *possible* for a genuine test of a theory, and, if this is the case, it seems that something has been gained if the test turns out to be positive. We then speak of the *confirmation* of a theory. As the geometric example

$$d(x^\circ, y^\circ) \neq 1, \quad d(y^\circ, z^\circ) \neq 2, \quad d(x^\circ, z^\circ) \neq 4 \tag{3.9}$$

shows, however, even this requirement can still be very weak. For (3.9) is compatible with (3.1) and has the variant

$$d(x^\circ, y^\circ) = 1, \quad d(y^\circ, z^\circ) = 2, \quad d(x^\circ, z^\circ) = 4 \tag{3.10}$$

that contradicts the triangle inequality. The inadequacy of this example is obviously due to the asymmetry between affirmation and negation of the distance statements: The specification of a distance contains infinitely more information than its negation. This is why (3.9) is so weak and will hardly be considered an empirical success of (3.1). This would be different if affirmation and negation contained roughly equal amounts of information. Although the contingent statements of physics usually exhibit this asymmetry, we do not want to neglect this case of a confirmation.

However, our genuine interest is the case in which, beyond the confirmation in (3.5), the theory induces an implication. This obviously does not yet happen in (3.9): None of these statements follows from one of the others, or the other two, together with the laws (3.1). However, this is certainly the case if the observations resulted, e.g., in:

$$d(x^\circ, y^\circ) = 1, \quad d(y^\circ, z^\circ) = 2, \quad d(x^\circ, z^\circ) \neq 4\,. \tag{3.11}$$

Here, at least in a summary, it is true that one of both statements follows from the others (in the present case, the third from the other two), if we add (3.1) or even just the triangle inequality. In such a case, we are entitled to call (3.11) an *empirical* success of (3.1). The success lies in the fact that we could have stated a contingent proposition as a *correct prediction* with the help of other contingent statements but, above all, *with the theory* without having already made a corresponding measurement or known its result. Hence, Hempel (1965, 26–31) here refers to a 'prediction criterion' of confirmation. In general, we speak of (3.5) as an empirical success or failure of the theory if (with Σ as its axioms)

$$\Sigma \wedge A_i(\ldots) \wedge \cdots \wedge A_k(\ldots) \vdash A_l(\ldots) \text{ respectively } \neg A_l(\ldots) \tag{3.12}$$

also holds in a certain order, in addition to the compatibility of (3.5) with Σ. Other examples besides (3.11) are:

$$x^\circ \neq y^\circ, \quad d(x^\circ, y^\circ) \neq 0$$

$$d(x^\circ, y^\circ), \quad d(y^\circ, x^\circ)$$

If they apply, they are obviously also empirical successes of (3.1). Note, here, that the distinction between successes and failures is not made solely on the basis of the logical relations of (3.5) to the theory, but always presupposes in addition that all the statements (3.5) *hold*. If this is not presupposed, a success can just as well be a failure, namely, if the inferred statement is false. In order to avoid misunderstandings, we speak of *possible* successes and *possible* failures as long as the statements (3.5) just satisfy the logical relation (3.12) to the axioms Σ, and of (actual) successes or failures if they additionally hold.

If there is an empirical success with regard to EG, then both the premises and the conclusion (according to 3.16) are about *one and the same* structural element, namely, d. The same is essentially true for the theory of the particle in the field with regard to its motion $\mathbf{x}$. More generally, we will have the case that in the premises (according to 3.16) statements of the form (3.10) are made about several structural elements s_i and, in the conclusion, about one of them. However, it also happens that we infer from the knowledge of some structural elements a statement about *another* structural element not yet mentioned in the premises. This is the case with theories that combine different (1-dimensional) quantities in a law in such a way that we can deduce the value of just one of these quantities from the values of all but one of them. An extremely simple case is the law

$$pv = RT \tag{3.13}$$

for ideal gases. Here, pressure, volume, and temperature are linked in such a way that each of these magnitudes is a function of the other two (with numerically given gas constant R). If in this case $p^\circ, v^\circ, T^\circ$ are three values that satisfy (3.13), then the set of statements

$$p = p^\circ, \ v = v^\circ, \ T = T^\circ \tag{3.14}$$

about the gas in question is an empirical success if they hold, and here the required inference can proceed from any one of these magnitudes. We are dealing with an inference to a different structural element in each case, because the analysis of the theory shows that we are dealing here with three principal base sets X, Y, Z for the values of pressure, volume and temperature, each of which is connected to $\mathbb{R}^+$ by a fixed scale (cf. the following section). The axiom of the theory is given solely by the typifications

$$\mathbf{p} \in X, \quad \mathbf{v} \in Y, \quad \mathbf{T} \in Z \tag{3.15}$$

and (3.13), where p, v, T are the numerical values of $\mathbf{p}$, $\mathbf{v}$, $\mathbf{T}$. The three statements (3.14) are, thus, statements about different structural elements. But because they remain connected in (3.13), it is possible to deduce the third from two of them.

Theories of the type (3.13-3.15), which we will discuss more generally in the next section, are well suited to shed light on our notion of theory, according to which a theory refers first of all to some, albeit arbitrary, system and only secondarily the situation is considered (according to Sect. 2.3, (2.69–2.70)) that arises with regard to a domain of application. This order of precedence is also maintained for the concept of corroboration. It is quite legitimate to see the corroboration of a theory also in the fact that it can be established in as large a domain of application as possible. On the basis of our conception of a theory, however, no theoretical problem arises from the fact that a theory that has proven itself in application to a system $\check{r}_1$ now also proves itself for a second system $\check{r}_2$, a third $\check{r}_3$, etc. Rather, the statement that a theory has proven itself in such and such a domain of physical systems is understood as the mere conjunction of its having proven for each of these systems, and only here does the problem arise as to what the corroboration for a single system is supposed to mean. A theory like (3.13-3.15) now sheds light on these relations because, here, one tends to think that data for an empirical corroboration will only be gathered if many triples (3.14) have been determined for just as many *different* systems, i.e., in this case gases in different states. One tends to think this *in contrast* to the cases of geometry and the particle in the force field, where one tends to be satisfied already with correspondingly many statements (3.2) or (3.4) about *one and the same* system. But, here, always the same statements about the same distance function and the same motion were meant, and it is only due to the higher complexity of these objects that it is possible to make an infinite number of independent statements about them but only a few about a gas. The multiplicity of the corroboration cases notwithstanding, the parallels between the theories here are to be drawn as just outlined. The parallel exists between (3.2), (3.4) (each in finite number), and (3.14), and the content of our notion of corroboration is entirely with the theory of the single system.

In one respect, however, an empirical data collection on several systems of one kind (apart from statistical surveys, of course) is particularly important. This is the case if terms occur in a theory, the empirical determination of which is carried out with recourse to this very theory—whether this is done for reasons of principle or for more practical reasons. As already indicated above (Sneed 1971), it has been argued that also, in an ultimate structure of physics, certain quantities can only be introduced by theory (as opposed to direct measurement), e.g., mass and force by Newton's mechanics, and that they can then, however, only be measured by using this theory. On the other hand, we know, e.g., from the practice of land surveying that EG is used to determine the distance of inconvenient objects, and has to be (precisely for practical reasons).

It is the occurrence of such *intra-theoretical magnitudes*, as one might call them, which usually requires empirical investigations on several systems, if the values of such magnitudes are to be determined. Let us take the van der Waals equation for a real gas

$$(p + a/v^2)(v - b) = RT$$

as an example. Here, the constants a and b (and indirectly also R) are microphysical. This allows, in principle, for their external determination, but is outside of the phenomenological level of the other quantities p, v, T. Nevertheless, one can, of course, determine their values using the van der Waals equation. Obviously, however, only *one* value triple $\langle p, v, T\rangle$ is not sufficient. You need at least three, which means that you have to bring the gas into different states and determine p, v, and T in each. We are, thus, dealing with a survey on several systems, and we have not ruled out that the contingent statements (3.5)–(3.7) can arise from several systems. As already mentioned, this procedure weakens the empirical content of the respective law containing intra-theoretical magnitudes, but this weakening is insignificant, and there is no reason to consider intra-theoretical magnitudes as less real than those that can be determined in a more direct way. On the basis of the hypothetico-deductive notion of corroboration, the empirical content of a theory lies mainly in the universal implications that derive from its empirical concepts and its axioms (3.12). A further difficulty arises from the possibility that such implications may not be available at all and that the empirical content, therefore, decreases to zero. Hempel (1965, 28) gives, e.g., the statement

$$\forall x . \forall y . Qxy . \rightarrow \exists z . Rxz .. \tag{3.16}$$

as one that cannot have any empirical successes or failures in the singular statements Qxy and Rxy. This means, of course, that the prediction criterion (as Hempel calls it) does not result in genuine empirical control for every theory. But to reject it as an explicate for a concept of confirmation, as Hempel does, is going too far as long as one has not also shown that the criterion has no *scientific* significance; that is, that it suffers a similar fate for important theories of physics as for the statement (3.16), which was specifically made for this purpose, or that it cannot be brought to bear in any physical theory at all. This is not the case, however, and the truth is in the middle. Accordingly, other philosophers of science have taken the anti-empiricist position that physical theories are empirically underdetermined and that we have to live with that (cf. Quine 1975).

Indeed, even for a theory as simple as (3.3)–(3.4), in order to arrive at a defensible empirical account of the particle, we must assume that the force field $\mathbf{f}$ is given in toto, which can hardly ever be an empirical datum, given the continuous dependence of the force on any given point in space. Much more at risk in this respect are the field theories, such as the theory of heat conduction dominated by the partial differential equation

$$\kappa \nabla^2 u = \partial u / \partial t \tag{3.17}$$

(with u the temperature as a function of space and time). For these, in order to arrive at empirical statements, we must accept initial conditions of the form

$$\forall \mathbf{x} \in S.\ u(0, \mathbf{x}) = g^\circ(\mathbf{x}). \tag{3.18}$$

(S is the space, g° a temperature distribution there.) On the other hand, this is indeed justifiable for certain simple cases. In the 1-dimensional case, for example, with $\kappa = 1$ and

$$g^\circ(x) = \begin{cases} 1 & \text{for } |x| \leq 1 \\ 0 & \text{for } |x| > 1\ , \end{cases} \tag{3.19}$$

we obtain the singular predictions

$$u\left(\frac{1}{4}, 0\right) = 0.8427, \quad u(1, 0) = 0.5205 \tag{3.20}$$

(cf. Baule 1962, 72). Thus, one sees that, in physics, the concept of hypothetico-deductive corroboration can be sufficiently brought to bear, if one is willing to make certain idealising concessions. On a general level, nothing can make this clearer than the class of *deterministic theories*, which is specific but widespread in physics.

Systems of statements (3.5)–(3.7), as we have based the concept of hypothetical-deductive corroboration on, generally give only partial descriptions of the physical system a theory is about. And we have already said that it may happen that even by using the theory, one cannot get beyond a given partial description, i.e., one cannot extend the theory up to new contingent statements by drawing conclusions. However, there is also the reverse case, in which one can derive a *complete* contingent description of the respective system from partial descriptions together with the theory. These are obviously the cases in which the idea of hypothetico-deductive corroboration celebrates its greatest triumphs and which must, therefore, be considered more closely here. The best-known cases are the deterministic theories in the narrower sense of physical theories, such as (3.3)–(3.4). Here, the temporal evolution of a system is described by an equation of motion the complete solutions of which are uniquely determined by initial and boundary conditions (Laplacian determinism). It is easy to see that this concerns cases in which partial descriptions permit unambiguous extensions to complete (contingent) descriptions.

A relevant and at the same time well-known case is our example theory (3.3)–(3.4) of the motion of a particle in a given force field. The differential equation (3.3), which represents the law of motion, satisfies (under very general conditions) an existence and uniqueness theorem: Given the initial data m°, α° , $\mathbf{x}^\circ$, and $\mathbf{v}^\circ$, there is (in a neighborhood of t°) exactly one motion $\mathbf{x}(t)$ that satisfies the law (3.3) and the initial conditions (3.4) (restricted to the affirmations). Therefore, any further question can be answered from the initial data. 'Is the particle at time t at location x?' can be decided unambiguously in principle. The answer is 'simply' read off from the uniquely determined solution. Of course, our question and the answer refer only to what happens *according to theory* (3.3). But these predictions can be compared with what actually happens, and the hypothetico-deductive corroboration is based on this comparison: If we have agreement, it is an empirical success because the theory has correctly predicted what will happen. Otherwise, it is a failure.

We know today from chaos theory that there is not much in strict determinism, for example, concerning Hamiltonian mechanics: The predictions have an enormously limited practical significance as soon as the initial conditions are relaxed by accepting deviations, as they inevitably occur due to measurement inaccuracies (for this and other pitfalls of determinism, see Earman 1986). It is, therefore, all the more important to realize that exact determinism in the narrower sense—temporal determinism—is only a special case of a more general determinism, according to which the laws of nature generally make the complete contingent descriptions of the systems governed by them redundant to a very high degree. Since our only empirical access to the system is through its contingent description, this means that general determinism simply consists in giving a maximum of empirical consequences to a theory. If we were to give up this principle, we would be giving up one of the core elements of theoretical physics: the extensive theoretical calculation based on lawful and contingent assumptions for the purpose of explaining phenomena and confirming theories.

Let us, therefore, realize that EG is also a deterministic theory in the following sense. If we initially have only an (n-dimensional) affine space, then with respect to the specification of $n+1$ points $O, P_1, ..., P_n$ with linearly independent vectors $\langle O, P_i \rangle$, the following existence and uniqueness theorem applies: There exists exactly one distance function d in such a way that d is a Euclidean metric compatible with the existing affine structure and satisfies the 'initial conditions':

$$d(O, P_i) = 1, \quad d(P_i, P_k) = \sqrt{2} \text{ for } i \neq k \, . \tag{3.21}$$

Every other distance statement $d(P, Q) = \alpha$ can, therefore, be either proved or disproved from (3.21). To appreciate this, it is useful to set up the following fiction: We give names to all points of the space and likewise to all possible distances. In principle, this can be done with a coordinate system or a scale. On this, we set up a huge catalogue of statements of the form (3.2) so that, for any two points x and y and any distance α, one of the two d-statements (3.2) is listed in the catalogue. With respect to logic and set theory, this catalogue is not subject to any further conditions (except for the equality axioms). In other words, all statements in the catalogue are logically independent, in the sense that no subset of them implies another of these statements. For this very reason, one cannot reduce the catalogue by purely logical means. This situation changes at once if all the theorems of EG are fulfilled in the catalogue, i.e., if the catalogue is, as the logicians say, a *diagram* (of a model) of EG (cf. Chang and Keisler 1973, 68–69). Then countless dependencies between its statements are induced and its total information becomes redundant—as the above metatheorem shows, indeed, highly redundant: Finitely many statements suffice to reconstruct the full description (as is also the case in the particle theory).

We see from this and the other examples—also the theory (3.13)–(3.15) is easily shown to be deterministic—that the hypothetico-deductive corroboration is expected to play a role at least for theories for which a certain existence and uniqueness theorem can be proven, which is a generalization of the corresponding theorems about systems of differential equations known from the dynamics of physics. In the following, we give an idea of this generalization, but for the sake of simplicity we restrict ourselves to the 'homogeneous' case in which the conclusion belongs to one of the structural elements that already appears in the premises. We then have a species of structure

$$\Sigma(\ldots s_i \ldots s_n), \tag{3.22}$$

in which some typified terms are distinguished as being affected by what follows. In the applications, we are concerned with empirical terms, i.e., such terms for which arbitrary statements of the form (3.6) can be empirically decided. The typification of these terms will then in most cases be linear in Pow. Some of the s_i can be defined terms, in which case the definitions are included in the axioms, i.e. they are in Σ. As 'initial data' we now add new constants $s_1^*, \ldots, s_i^*$, which are subject to certain minimum conditions

$$\Sigma^*(\ldots s_i^* \ldots s_n^*) , \tag{3.23}$$

but at least to the condition that s_i^* has the same type as s_i (for all i). These conditions can affect the single s_i, e.g., differentiability, but they can also connect the s_i, e.g., when the position and velocity of a particle are specified for the same time. Above all, however, Σ^* must ensure that the set of initial data is compatible with the solutions of Σ. For we have

$$s_i^* \subseteq s_i . \tag{3.24}$$

as initial conditions. The reduction condition thus amounts to the subset relation: One only needs to have the subset s_i^* of s_i, then s_i is known by means of the theory. More precisely:

$$\Sigma^*(\ldots s_1^* \ldots s_n^*) \to \exists! s_1 \ldots s_n . \Sigma(\ldots s_1 \ldots s_n) \wedge s_1^* \subseteq s_1 \wedge \cdots \wedge s_n^* \subseteq s_n . \tag{3.25}$$

According to the examples presented, this theorem is the basis for the proof of every singular statement (3.6) about one of the s_i from (3.23-3.25)—the proof or refutation, depending on whether the statement in question is compatible with the premises or not. The theorem (3.26) emphasises, once again, that a theory is not deterministic per se, but with respect to a well-defined set of initial data, which can often be chosen in different ways for one and the same theory—just think of the canonical transformations in Hamiltonian mechanics.

Finally, it should be mentioned that the logical mechanism of inferring other contingent statements from laws together with contingent statements has another interpretation well-known just to physicists. So far, we have assumed that a theory is still unconfirmed (has not yet been empirically corroborated), and have asked for situations that can be regarded as a confirmation or invalidation, respectively. This logical situation comes into consideration if it has further been shown that all contingent statements in play are empirically valid. The inferred statement can then be considered a correct prediction made by means of the theory and, thus, an empirical success of the latter. Under somewhat different circumstances, however, the process can also be regarded as a (singular) *explanation* of that statement. The so-called D-N explanation, particularly investigated by Hempel and Oppenheim, referred mainly to cases in which the explanandum is not itself a theory but a simple singular statement (cf. Sect. 1.2). This entails that some of the premises are also contingent, while the dignity of the explanation is, of course, that *also laws* belong to the premises (cf. Hempel 1965, Sects. 10 and 12). However, this logical scheme can only be assessed as an explanation if the theory on which the explanation is based is empirically well confirmed: A poison murder is not clarified by finding in the body of the dead person a substance the digestibility of which has never been tested. In such a situation, this case can at best be regarded as a *confirmation* of the toxicity in question. The difference between a singular explanation and a hypothetico-deductive confirmation, thus, lies in the fact that, for the former, it is required as already done what the latter should first achieve: the confirmation of the assumptions of the theory. Since the corroboration of a theory is in practice never complete, the boundaries between this kind of explanation and confirmation are blurred.

3.2 Consideration of Measurement Inaccuracies

The corroboration theory outlined in the previous section is still insufficient because it does not take into account the phenomenon of measurement inaccuracy. The observation statements in (3.6) directly specify a structural element s of the physical system under investigation by an element $\alpha \in s$ that also directly belongs to this system. Thus, for example, for the metric d of Euclidean geometry or for a motion s of the particle of our example theory in Sect. 2.1, we arrive at observation statements of the form

$$d(p, q) = \alpha, \quad \langle \mathbf{x}, t \rangle \in s \,, \tag{3.26}$$

which state that α is the distance of the points p and q or that the particle (in the course of its motion) is at the position $\mathbf{x}$ at time t, respectively. Strictly speaking, however, we are not in a position to produce such statements by means of measurements. Measurements in the continuum, as prevailing in physics, are never absolutely exact: The position of the points p and q and the value of α in a distance measurement are just as little known exactly as the position and time in determining the motion of a particle. This circumstance looks like a principal deficiency of physics. But it is easy to become convinced that without it our physics would not have emerged at all—at least not theoretical physics. For we know that every theory has been eventually falsified by sufficiently accurate measurements, and we know that if theories are wrong anywhere, they are usually wrong throughout. So, not only are there eventually falsifying measurements, but indeed any sufficiently accurate measurement would falsify. If we were able to measure with absolute precision, the lifetime of physical theories would be very short. Our weakness in not being able to measure with absolute precision compensates in a wonderful way for our other weakness in not immediately arriving at the final theory. Hilbert (1919, 52–53) remarked on this:

> The laws [...] are not found all at once [...]. It is the [...] inaccuracy of our observations which comes to the fore here. On the one hand, this is a sorry state of affairs; on the other hand, it also helps us to find the laws. If we had perfectly accurate observations, we would be completely overwhelmed by the multitude and complication of phenomena [...]. Thus, for example—as Poincaré once remarked—Kepler would not have arrived at his laws if he had observed with complete accuracy.

This situation has consequences for a study that is primarily concerned with theory reduction. As already indicated in the introduction, in the further course of our investigation, we want to understand, among other things, that theory reduction makes empirical progress possible. As long as it is a matter of exact reductions, we can get by with the notion of exact corroboration, i.e., corroboration based on exact observation statements when it comes to defining the notion of empirical progress (see the following section). More important than the exact reductions, however, are the approximate ones. The interesting theory reductions are those in which a relatively large error is involved, so that the new theory T', while improving on its predecessor T, is somehow incompatible with it. If one were to rely on a concept of corroboration that only corresponds to exact measurements, such as those developed in the previous section, then *no* empirical success of T would also be a success of T'. If T' were then 'absolutely correct', then those possible successes of T could not occur at all, and there would not only seem to be no reason to save them at least as possible (if not actual) successes in the new theory, but this saving would be completely undesirable. Scientific reality, however, looks different. On the basis of a concept of corroboration that takes measurement inaccuracies into account and is therefore liberal, T and T' *can* have common approximate empirical successes despite their incompatibility. And, on the basis of a corresponding notion of approximate corroboration, approximate reductions also find an empirical justification (cf. Chap. 5). It is to Günther Ludwig's credit to have emphatically drawn attention to the significance of measurement inaccuracies also in theoretical considerations on physics (cf. Ludwig 1990, Sect. 6; cf. also Girill 1978).

To obtain a notion of approximate corroboration motivated by the possibility of approximate reductions, we must first return to the observation statements (3) of the previous section, which function as the possible outcome of a test, and weaken them to *imprecise observation statements*. For the purpose of illustration, we will introduce a class of theories, the so-called simple theories, for which a gas equation, Kepler's 3rd law, Snell's law of refraction, Ohm's law, etc., are examples. It is, therefore, a matter of theories whose respective law expresses a relationship between a finite number of physical quantities and can accordingly be characterized by a relation between numerical quantities—a number relation. In terms of our conception of a theory from 2.3, the species of structure of a simple theory would have to be described as follows.

We have n principal base sets $X_1, \ldots, X_n$, which are simply the domains of n magnitudes measurable on a system. Correspondingly, we have n scales $\phi_1, \ldots, \phi_n$, with the typification

$$\phi_i \subseteq X_i \times \mathbb{R} , \tag{3.27}$$

of which we require that they uniquely map the X_i to (possibly infinite) intervals I_i of real numbers:

$$\phi_i : X_i \to I_i \quad (I_i \subseteq \mathbb{R}) . \tag{3.28}$$

Finally, we have n variables $x_1, \ldots, x_n$ with

$$x_i \in X_i , \tag{3.29}$$

which are subject to a law

$$\langle \phi_1(x_1) \ldots \phi_n(n) \rangle \in \rho \,, \tag{3.30}$$

where

$$\rho \subseteq I_1 \times \cdots \times I_n \subseteq \mathbb{R}^n \tag{3.31}$$

is a fixed numerical relation, which will usually be an ordinary hyperplane in $\mathbb{R}^n$ and can often be uniquely resolved with respect to each variable. ρ can depend on further constants of the theory, but for simplicity we think of them as numerically fixed. In the examples just given, the gas constant, the mass of the sun, the refractive index, the electrical resistance, etc. are such constants.

For simple theories, the observation statements corresponding to (3.1) and, more generally, (3.5)–(3.7) of the previous section are of the form

$$\phi(x) = a \,, \tag{3.32}$$

where we suppress the index i for the sake of simplicity. Here, a is the measured *numerical* value of the magnitude X in question, and (3.32) is the statement that its value x in the description is numerically just a. Except ϕ, then, observational data and statements of simple theories are of 1st order. Given a full set of statements (3.32) for all magnitudes $X_1, \ldots, X_n$, but still for only *one* system, it is clear why the simple theories make life easy in terms of corroboration: Every solution of the law (3.30) can be *completely* described by such a finite conjunction of observation statements. This advantage will later also become apparent in the case of approximate corroboration.

Now, the basic idea for replacing (3.32) by an approximation is to say, instead of (3.32), only that x has a numerical value a' which is "close to" a, whereby the "close to" would have to express that one can no longer distinguish by measurement between a and a'. But what "close to" itself means must be clarified on a topological basis. Here, we follow Ludwig's suggestion not to use simply a topology on X or $I = \phi(X)$ but a uniform structure for the definition (Ludwig 1990, Sect. 6). This has the advantage that the definition of the neighbourhood of a (and thus of the values that can no longer be distinguished from a within the measurement inaccuracy) does not depend on a. The question remains, of course, which uniform structure to choose. First of all, the ordinary metric on v suggests itself, which would lead to replacing (3.32) by

$$\text{there exists } a' \in I, \text{ such that } |a - a'| < \epsilon \text{ and } \phi(x) = a' \,, \tag{3.33}$$

where ϵ would reflect the measurement inaccuracy. However, this choice would usually be rather inappropriate because it measures the quality of a measurement by the absolute error. In the case of a mass determination, for example, an error of one gram would be extremely small for a measurement of the mass of the Earth, i.e., the measurement would be very accurate, whereas a measurement of the mass of a hydrogen atom with the same error would be absolutely useless. We get an improvement of this situation if we use the relative error

$$|a - a'|/a < \epsilon \tag{3.34}$$

instead of the absolute error in (3.33). But, of course, the actual measurement situation may require other uniform structures on I.

This brings us to the question of what the general situation is. With regard to the *uniform structure*, we generally have neighbourhoods u with $u \subseteq I^2$, so that (3.33) becomes

$$\text{there exists } a' \in I, \text{ such that } \langle a, a' \rangle \in u \text{ and } \phi(x) = a' \,. \tag{3.35}$$

For the conditions that the u have to meet in order to constitute a uniform structure, we must refer to the special literature (cf. Querenburg 1973, Chap. 11; Bourbaki 1965, Chap. 2). On the other hand, as far as the present context is concerned, an equation like (3.32) usually applies only to cases where we are dealing with elements of principal base sets the own elements of which (if any) no longer have any physical meaning within the system to be described (just as in (3.29)). Already in our example theory of Sect. 2.1, we meet other conditions in this respect. Let us consider the motion b of the particle in space (S) and time (T). Then, we have as a starting point the typification

$$b \subseteq T \times S \tag{3.36}$$

and, as exact observation statements about the motion, the statements

$$\langle t^\circ, \mathbf{x}^\circ \rangle \in b \,. \tag{3.37}$$

If we want to take the measurement inaccuracy into account here, we would have to introduce a uniform structure Un on $T \times X$, with which we then, after specifying a measurement inaccuracy $u \in Un$, weaken the statements (3.37) to the effect that the time t° and the position $\mathbf{x}^\circ$ themselves no longer belong to the motion, but only a time t and a position $\mathbf{x}$ having no larger distance than u from it:

$$\begin{aligned} &\text{there exists } t \in T, \mathbf{x} \in S, \\ &\qquad \text{such that } \langle \langle t^\circ, \mathbf{x}^\circ \rangle \langle t, \mathbf{x} \rangle \rangle \in u \text{ and } \langle t, \mathbf{x} \rangle \in b. \end{aligned} \tag{3.38}$$

In general, we are dealing with observation statements of the form (3.6) and (3.7) of the previous section. For (3.6), we have a typification

$$s \subseteq \sigma_1(X_1^\circ) \times \cdots \times \sigma_n(X_n^\circ) \,, \tag{3.39}$$

and for (3.7) usually

$$s_i \in X_i^\circ \,. \tag{3.40}$$

The composition of the σ_i is not important in the following. In general, we will also deal with principal basis terms. It must be explicitly noted, however, that we do not deal here with the case in which the typification of s would be given by $s \in \mathrm{Pow}^2 \ldots$. For the approximate weakening of the statements (3.6) of the previous section, we need a uniform structure on the $\sigma_i(X_i^\circ)$ (for the equations) and on their product (for the membership statements). For the latter, we will take the uniform product structure of the former (cf. Querenburg 1973, 11.24). To weaken the statements (3.7) of the previous section, a uniform structure on each of the X_i° must be given. So, for the following, we assume that in both cases a uniform structure is *deducible* from the core structure of the theory together with its axioms. (Note that this deduction, of course, must not depend on the variable structural elements s. Rather, as variables, these are each subject to one and the same uniform structure, which is deducible from the *core* of the theory.)

$$\exists a'.\langle a, a' \rangle \in u \wedge a' \in s.. \tag{3.41}$$

With these specifications, we now obtain as an approximate version of (3.6), i.e., in particular for 2nd-order observation statements, with given imprecision u, for $a \in s$ (with the abbreviation a for $\langle a_1, \ldots, a_n \rangle$ and the renaming a_k for x_k°). For (3.41), one can also write

$$a \in s_u\ , \tag{3.42}$$

if one generally posits

$$M_u = \{x | \exists y.\langle x, y \rangle \in u \wedge y \in M.\} \tag{3.43}$$

for a subset M of the uniform space. Correspondingly, for the equations $a_i = a_i'$ in (3.6) one obtains

$$\exists x.\langle a_i, x \rangle \in u \wedge x = a_i'.\ , \tag{3.44}$$

which is, however, equivalent to

$$\langle a_i, a_i' \rangle \in u\ . \tag{3.45}$$

Therefore, for the Eq. (3.7), i.e., for 1st-order observation statements, on their space one also obtains

$$\langle s_i, a_i \rangle \in u \tag{3.46}$$

as their version taking into account the measurement inaccuracy. The negated statements of (3.6)–(3.7) receive as their equivalents the negations of (3.41), (3.44), (3.46). It should be noted, however, that while the affirmations $a \in s$ are weakened by the transition from s to s_u, the negations $\neg a \in s$ are *strengthened* to $\neg a \in s_u$, and the same applies to the equations. This is not a misfortune, however. As has already been noted, the exact negations are, in general, extremely weak relative to the affirmations they negate and, so, this discrepancy is somewhat compensated for when the measurement inaccuracy is taken into account.

Equipped with these tools, we can now address the important question of what constitutes an *approximate corroboration* as opposed to an exact one. We do not aim at full generality. Let the possible outcome of an approximate test

$$\cdots \wedge a \in s_u \wedge \cdots \wedge b \in t_v \wedge \cdots \tag{3.47}$$

be an approximate observation report, where the conjunctions are weakenings of corresponding exact statements according to (3.41), (3.44), (3.46), and negations may also occur. Then, it is obvious to consider (3.47) as a *possible approximate confirmation* of the species of structure

$$\Sigma(\ldots s \ldots t \ldots) \tag{3.48}$$

in the context of the measurement inaccuracies $\ldots u \ldots v \ldots$ if

$$\vdash \ldots \exists s \ldots \exists t \ldots \Sigma(\ldots s \ldots t \ldots) \wedge \ldots a \in s_u \ldots b \in t_v \ldots \tag{3.49}$$

holds. (Note that according to (3.43) s_u and t_v are functions of s and t respectively.) However, as will be shown in a moment, it is better to require

$$\begin{aligned} &\ldots \exists a' \ldots \exists b' \ldots \langle a, a' \rangle \in u \wedge \ldots \langle b, b' \rangle \in v \wedge \ldots \\ &\quad \wedge \ldots \exists s \ldots \exists t \ldots \Sigma(\ldots s \ldots t \ldots) \wedge \ldots a' \in s \wedge \ldots b' = t \ldots \end{aligned} \tag{3.50}$$

instead, for reasons of consistency. Equation (3.49) directly states that Σ is compatible with the approximate observation report (3.47). Equation (3.50), on the other hand, states that with certain data close to the data of (3.47), the exact observation report

$$\ldots \wedge a' \in s \wedge \ldots b' = t \wedge \ldots$$

is compatible with Σ. Thus, one could also say that (3.49) provides the exact compatibility of the approximate observation report, whereas (3.50) provides the approximate compatibility of the exact report. This interchange also corresponds to the easily given proof that the two formulations are equivalent. So it does not matter here in the end which of the two formulas we take to be the definition of the notion of approximate confirmation. The same then applies to the notion of *possible approximate failure*, which is defined by the negation of (3.49), or, alternatively, of (3.50). Compared to the exact relationships, then, approximate confirmation is more accommodating than exact confirmation, while the reverse is true for failures.

The situation is somewhat different for the notion of *possible approximate success*. If we were to choose the form corresponding to (3.49) here, we would obtain

$$\vdash \ldots \forall s \ldots \forall t \ldots \forall r \ldots \Sigma(\ldots s \ldots t \ldots r \ldots) \wedge \ldots \\ a \in s_u \wedge \ldots b \in t_v \wedge \ldots \rightarrow c \in r_w \ldots , \tag{3.51}$$

where r may (but need not) coincide with one of the terms in the premise, and negations (instead of affirmations) may occur. For (3.51), however, the case can arise that in favour of the possibility of inferring the conclusion from the premises 'weakened' by the measurement inaccuracies, one must choose their measurement inaccuracy w unnecessarily large—larger than the available measurement possibilities would allow. This teaches us the chaotic behaviour of many dynamic systems that move according to classical mechanics: With only very small deviations in the initial data, these can become so large over time that a prediction becomes practically useless because of a too large w in r_w, however small $u \ldots v \ldots$ are chosen. By means of simple examples, it is easy to see how the applicability of (3.51) depends on the uniform structure on which it is based. If, for example, for Boyle's/Mariotte's law $pv = c$ with fixed c, we choose the natural metric according to (3.33) for v as well as for p, then, because of

$$|p - p^\circ| < c/(v \cdot v^\circ) \cdot \epsilon \text{ for } |v - v^\circ| < \epsilon ,$$

we get very large deviations for p if v is small for any given accuracy of measurement for v. If, on the other hand, we choose the uniform structure of the relative error according to (3.34), then because of

$$|p - p^\circ|/p^\circ = |v - v^\circ|/v^\circ$$

the measurement accuracy remains essentially the same. We get a little more independence from these problems if we choose the formulation

$$\ldots \exists a' \ldots \exists b' \ldots \exists c' \ldots \langle a, a' \rangle \in u \wedge \ldots \langle b, b' \rangle \in v \wedge \ldots \langle c, c' \rangle \in w \\ \wedge \ldots \forall s \ldots \forall t \ldots \forall r . \Sigma(\ldots s \ldots t \ldots r) \wedge \ldots a' \in s \wedge \ldots b' = t \\ \wedge \ldots \rightarrow c' \in r , \tag{3.52}$$

corresponding to (3.50) for the success case, instead of (3.51). Here, instead of the compatibility, it is the deducibility which is not required for $a \ldots b \ldots c$ itself, but for $a' \ldots b' \ldots c'$ that cannot be distinguished from them within the measurement inaccuracy. It is easy to see that under self-evident compatibility conditions (3.52) follows from (3.51). However, the reverse is not true, as can be seen by choosing sufficiently small w for (3.52).

3.3 Empirical Progress

In this section, we are concerned with the question of under which conditions we are justified to say that a theory T' *is* an empirical advance over a theory T, and also under which it *can be*. In dealing with these questions, we introduce a subject which will occupy us again and again throughout the further course of our investigation: the relationship between progress and reduction, as well as that between theoretical and empirical reduction and corresponding progress. It is easy to conjecture about these relationships that theoretical reduction—reduction of theories—makes empirical progress *possible*: If we have a reduction of theory T to theory T', then T' *can* be an empirical advance over T. More, of course, is not: Whether T' *actually is* an advance over T depends on the actual situation of corroboration, and for that (as being absolute), the reduction (as being relative by its nature) does not provide the slightest hint. But the possibility of progress could be given by the reduction, and that would be more than nothing. For it is possible to think of conditions under which progress is impossible.

Above all, however, this connection would be important because with the concept of empirical progress, and *only* with it, one might have found a necessary concomitant of theoretical development in general. For reduction alone may be by no means necessary for this. It may well be that in the development of physics such great leaps, not to say mistakes, are made that a reduction of the relevant theories is out of the question, but at least empirical progress has been made. The inversion of the relation

$$\text{theory reduction} \vdash \text{possibility of empirical progress} \tag{3.53}$$

does not hold at all, and in Chap. 4 we will even be forced to insert an intermediate link between the two partners in (3.53) with so-called partial reductions. But we want to consider (3.53) as a kind of adequacy condition for the concept of reduction, meaning that every reduction (of whatever kind) from T to T', in a sense to be defined, enables an empirical progress of T' relative to T. If, in the case that T' is an empirical advance over T, one also speaks of the fact that then T is empirically reduced to T', and then (3.53) can also be expressed by

$$\text{theory reduction} \vdash \text{possibility of empirical reduction,} \tag{3.54}$$

where on the right a limiting concept of reduction is meant.

In order to fix our concept, one must of course have a corroboration theory at one's disposal. For this very purpose, we have indicated, in the two previous sections, the theory of hypothetico-deductive corroboration sometimes as exact, but then also as approximate corroboration. At the beginning of this chapter, we also briefly considered the question of whether the introduction of an approximate concept of corroboration would not, in consequence, entail the physical theories themselves being conceived only approximatively. We had answered this question in the negative for this work. (This is an important point of difference to Ludwig 1990, cf. the appendix to the second part.)[1] However, the situation is different for the reductions. In the following, we will significantly distinguish between exact and approximate theory reductions. The consideration of approximate reductions is primarily theoretically motivated. In order to give the concept of reduction the weight it deserves for historical reasons of theory formation, we must also allow for reductions in which theories are reduced to other theories that are incompatible with the former. Such reductions, however, are necessarily approximate (cf. Chap. 5). But approximate theory reduction is also motivated by the fact that, with an approximate concept of corroboration, we will automatically also have a concept of approximate *empirical* reduction. And this concept would be redundant if it were not accompanied by a concept of approximate theory reduction. We will, thus, have to distinguish between the implication

$$\begin{aligned}&\text{exact theory reduction} \vdash \\ &\qquad \text{possibility of exact empirical reduction}\,,\end{aligned} \tag{3.55}$$

on the one hand, and

$$\begin{aligned}&\text{approximate theory reduction} \vdash \\ &\qquad \text{possibility of approximate empirical reduction}\end{aligned} \tag{3.56}$$

on the other, thus splitting our original objective (3.53)–(3.54) into these two cases. The possibility in (3.55) naturally includes the possibility in (3.56), while the reverse does not hold. An exact theory reduction can also be empirically justified approximatively, but an approximate one cannot be justified exactly.

After this differentiation of (3.53)–(3.54), we now have to note that our goal indicated by (3.53)–(3.54) cannot be achieved at all with regard to the concept of corroboration to be applied if, as done in the previous two sections, this concept is only used for the elementary statements (3.6)–(3.7) and (3.41), (3.44), (3.46) based on the basic concepts of a theory. Here, one could already express the internal criticism that there are also defined concepts as well as statements in a theory (cf. Sect. 2.4), to which the concept of hypothetico-deductive corroboration could be extended without further ado. Actually, in many cases there are *alternative* systems of basic concepts and basic statements (cf. Sect. 4.2), which (formally considered) could just as well have been taken as a starting point. The question is, therefore, why the material thus indicated is not used for possible corroborations of the theory. The answer, however, results precisely from our objective (3.53)–(3.54). It is precisely

[1] Scheibe planned to add to Vol. II an appendix about Ludwig's approach, which however was finally not realized; see Table of Contents Vol. II at the end of this book. Scheibe (1982a) compares the approaches of Ludwig (1990) and Sneed (1971). In an unpublished lecture, Scheibe (1987un, Ch. VIII) explained Ludwig's approach in more detail.

the reductions that lead us to ever other possibilities as indicated, i.e., extensions of the material of corroboration, and we would be anticipating our very subject if we wanted to exhaust all possibilities of empirical progress at this early stage. In addition, according to our explanations of Sect. 1.3, the topic of theory reduction is substantially open, and we cannot deal with it with a closed notion of reduction. Thus, we can only get to know the cases of corroboration resulting from reduction piecemeal.

For a theory T' to be an empirical improvement of a theory T, one must first ensure that it is impossible for it to be a deterioration. In this sense, we have already heard the admonition of Boltzmann and Nernst on the part of physicists that even a theory (T) that has been overcome can have had its merits within certain limits (cf. Sect. 1.1). The philosophers, especially Popper, also made it known that progress, even if it should not be purely cumulative, has a *conservative component* without which its asymmetry would not be guaranteed (cf. Sect. 1.2). And, without this asymmetry, which does not tolerate deterioration alongside improvement, still speaking of progress would be an abuse of language. Thus, we first have to formulate certain *conservation principles* that exclude deterioration. However, it must also be taken into account from the outset that, in the case of progress that is not purely cumulative, the conservation principles can only be relatively weak (cf. 3.56): If we want to speak of a possible progress even if T and T' contradict each other, then by no means all—indeed, in general, *none*—of the possible phenomena exactly explained by T can also be exactly explainable by T'. Rather, the possibility of progress will then rely significantly on the fact on the fact that only sufficiently imprecise explanations by T are also possible explanations by T'.

As a first conservation principle, let us consider that the domain of application $\mathfrak{N}$ of T as the theory to be improved is contained in the domain of application $\mathfrak{N}'$ of T'. For the focus of our conservative demands will be that (at least certain) possible successes of T are also successes of T'. A success of T, however, is always achieved on systems $y \in \mathfrak{N}$ only. Therefore, as a success also of T', $y \in \mathfrak{N}'$ must also apply to it:

C_1: The domain $\mathfrak{N}$ of T lies in the domain $\mathfrak{N}'$ of T'.

It has already been noted earlier (Sect. 2.3) that the concept of the domain of application is an obscure concept which we try to get over as harmlessly as possible. In any case, an application domain is not already the scope of the theory in question and is, therefore, not destined to be connected with it. Accordingly, C_1 is to be understood in such a way that we should choose the application domain of T' in such a way that it contains the domain of application intended for T.

Next, we come to the tests of T and T'. With them, we have already reached the point where we are taught that satisfying our conservation principles requires a certain liberality for the terms of the corroboration theory invoked. Are tests of T also always tests of T' if we want empirical progress? This seems desirable since the result of a test of T can be, after all, a success of T. And, again, as with the domains of application, we will conclude further: If this success can also be a success of T', then the same must apply a fortiori to the test. It, therefore, seems desirable to demand:

C_2: Every test of T is also a test of T'.

But now comes the following consideration, which takes up our general warning about wanting to establish a concept of corroboration independently of the expected theory reductions. It can happen that, within a theory, we are confronted with tests the generating statements of which do not belong to the basic statements of the theory. Such a case would be a test of Kepler's 3rd law within the full Keplerian theory. In this test, we would be faced with the question of whether a given length a and duration T, as the length of the major principal axis and orbital period of a Keplerian orbit, are related to each other in the way that Kepler's 3rd law asserts. This is undoubtedly a test of Kepler's theory, i.e., of Kepler's three laws of planetary motion. But it would not be a direct test of statements about when the planet is at which place, i.e., it would not be a direct test of the full Keplerian theory, but only of a coarsening of it, as it exists with Kepler's 3rd law (cf. Sect. 4.4).

With C_1 and C_2, we are still in the preliminary area of the really important conservation postulates. As presuppositions of the latter, which they actually are, we have split them off from the latter. When we now come to the postulates for confirmations and successes of theories, we must first of all bear in mind what we have already stated at the beginning of the previous section: If the theories T and T' contradict each other, there is, anyway, no hope of preserving absolutely exact confirmations or successes. Instead, we will will only be able to have something like

C_3: Every exact or not too exact confirmation of T
is a corresponding confirmation of T'.

And correspondingly:

C_4: Every exact or not too exact success of T
is a corresponding one of T'.

Here, we distinguish between the two main cases of exact and approximate empirical reduction. The inaccuracy required for the approximate case depends on the data of the confirmation or success with regard to the approximation of T to T'. Where this already goes very far, quite accurate measurements can already lead to confirmations or successes of both theories; where, on the other hand, the theories diverge roughly, they also have only very rough results in common.

To avoid misunderstandings of the conservation principles C_3 and C_4, a somewhat subtle point must be made that has nothing to do with the approximation relationships. In both cases, C_3 as well as C_4, we mean by confirmations and successes, respectively, possible confirmations or successes and distinguish these from the actual confirmations or successes, which, in contrast to the former, require the truth of the underlying observation statements (cf. the two previous sections). The possible, but not the actual, confirmations and successes depend solely on the respective theory. Against the conservation postulates C_3 and C_4, it could be objected that it is completely sufficient to require the conservation of the actual confirmations and successes, while one may be completely indifferent to those that are possible but have not occurred. This argument could be extended by the following consideration: Let us take, say, C_4 and a possible success $\langle A; a \rangle$, where A summarizes the premises that are compatible with the theory in question, T, and a is the conclusion from the axioms of T and A. Now, if $\langle A; a \rangle$ is a possible success of T, then $\langle A; \neg a \rangle$ is a possible failure of T, and the same is true with respect to the reducing theory T'. Thus, if we require C_4, then with every possible success the corresponding possible failure of T is also inherited by T', and it is questionable whether this should be brought about without necessity. For actual failures are also transmitted in this way. But, if we restricted ourselves in C_4 to actual successes, nothing at all follows about the preservation of actual failures. For if $\langle A; a \rangle$ is an actual success of T, then $\langle A; \neg a \rangle$ is still a possible, but certainly not an actual failure—neither of T nor of T'.

At first glance, this argument seems irrefutable and, yet, the following can be said against it. First of all, as far as the implication of possible failures is concerned, it should be pointed out that, in general, not all possible failures are inherited, despite C_4. For not all failures of T are of the form $\langle A; \neg a \rangle$ with possible success $\langle A; a \rangle$. For this to be the case, A would have to be compatible with T. $\langle A; \neg a \rangle$ (and also $\langle A; a \rangle$) can, however, be a possible failure of T due to the fact that A *alone* contradicts T. As a relevant case, we will later (Sect. 4.1 (4.28)) get to know the Newtonian generalisation (T') of Kepler's planetary theory (T)—the direct generalisation is meant, in which only the parabola and hyperbolic trajectories are added. If we take here $\langle A; a \rangle$ as a possible success of T', for which A are the initial conditions of a hyperbolic trajectory, then A is incompatible with T, hence $\langle A; a \rangle$ is a failure of T. In this way, a failure of T may well turn (literally) into a success of T'. Moreover, it should be noted, in general, that we are aiming at a notion of empirical reduction which, in the sense of (3.53)–(3.54), *guarantees* that every theoretical reduction entails an empirical one. This makes the above conservation principles *subjects of proof*, the conclusions of which require sufficient reasons to be found in the respective theoretical reductions. These reasons, however, can never substantially concern the factuality of confirmations or successes, but only their logical connection with the theories T and T'. In other words, if one can prove, e.g., C_4 for actual successes under the condition of a theoretical reduction, then this proof will not be able to make use of the factuality at all, but will always run via a proof of C_4 with only possible successes. In a reduction-theoretical context, therefore, this stronger claim must be required.

From the conservation principles, let us now move on to the *empirical improvements* that a theory T' can have over a theory T, if, moreover, C_1 to C_4 are already assumed. If one summarizes their validity to the effect that T' cannot be worse than T, then one easily slips into the obvious question of whether T' is really in fact better than T. However, this is not what we are concerned with at first; rather, consequently, we must first ask whether T' *can* be empirically better than T. Even this question goes beyond what has been dealt with so far, although it also only asks about the possibility of an improvement. The questions as to whether T' is *in fact* better or worse than T should be distinguished from it. First of all, however, we are dealing with something like:

> *Im*. The theory T' has (possible) confirmations and (possible) successes that are not confirmations or successes of T.

Naturally, such a mere existential requirement is relatively weak when compared to C_3 or C_4. But, in general, nothing can be effectively done about it, just as in many similar cases. We must let the specific cases determine the form of the positive success surplus of T' over T. After all, I is not empty, as can be seen, for example, in the quite innocent case that T' and T are equivalent (cf. Sect. 4.2). In this case, all conservation principles are still fulfilled, but not Im. In contrast to C_3 and C_4, we will not take up Im in the following. For the individual types of reduction, satisfying this postulate basically means that the reduction in question is not a trivial example of its kind. And there is little point in always explicitly ruling this out.

With regard to the actual conditions, there is a clear difference between C_3 and C_4, on the one hand, and Im on the other. Whereas, from the former follow their actual counterparts, this does not hold for Im. If possible successes of T are also possible successes of T', then actual successes of T are also actual successes of T'. But, from the fact that there are possible successes of T' which are not successes of T, it does not follow that there are actual successes of T'. It only follows that, if there are any, which are not possible successes of T, then they are not actual successes of T either. Accordingly, in the relevant context, one repeatedly finds this requirement—in our terminology:

> *Im'*. There are actual confirmations and successes of T' that are not (possible) confirmations or successes of T'.

In other respects, too, little can be said about the actual conditions and, certainly, nothing about the actual corroboration of T and T' taken separately. But, even about the relative circumstances, it only follows that a good situation of corroboration at T' yields the same for T, if T is reducible to T'. In particular, both theories can totally fail empirically, even if they are in a reduction relation, and it is possible that, under the same condition, with empirically good T, the theory T' may totally fail. For a justifiable empirical reduction, therefore, there remains to be required (cf. Nagel 1961, 358–366):

> A': The reducing theory T' is well approved.

But like Im', and unlike C_1 to C_4 and Im, this is not a statement resulting from a theoretical reduction relation of the theories in question.

References

Baule, B. 1962. *Die Mathematik des Naturforschers und Ingenieurs, vol. VI: Partielle Differentialgleichungen*. Leipzig: Hirzel.

Bourbaki, N. ^{4}1965. *Éléments de Mathématique, Livre III: Topologie Générale*, Ch. 1 et 2, Paris: Hermann.

Carnap, R. 1958. Beobachtungssprache und theoretische Sprache. *Dialectica 12, 236–247.* Engl. transl.: Observational Language and Theoretical Language. In: J. Hintikka (Ed.), 1975, Rudolf Carnap, *Logical Empiricist: Materials and Perspectives*, Dordrecht: Springer Netherlands 1975.

Chang, C.C., and H. J. Keisler. 1973. *Model Theory*. Amsterdam: North-Holland.

Earman, J. 1986. *A Primer on Determinism*. Dordrecht: Reidel.

Girill, T.R. 1978. Approximative Explanation. In: P. D. Asquith and I. Hacking (eds.), *PSA 1978*, vol. 1. East Lansing, Mich., 186–196.

Hempel, C. G. 1965. *Aspects of Scientific Explanation and Other Essays in the Philosophy of Science*. New York: The Free Press.

Hilbert, D. 1919. *Natur und mathematisches Erkennen. Lectures Gottingen 1919-1920.* From notes by P. Bemays. Ed. by D. E. Rowe. Springer: Basel 1992.

Lenzen, W. 1974. *Theorien der Bestatigung wissenschaftlicher Hypothesen*. Stuttgart: Frommann-Holzboog.

Ludwig, G. 1990. *Die Grundstrukturen einer physikalischen Theorie*. Berlin: Springer.

Nagel, E. 1961. *The Structure of Science*. New York: Harcourt, Brace and World.

Popper, K. 1935. *Logik der Forschung*. Wien: Springer. Tübingen: Mohr. Quoted after the Engl. transl.: *The Logic of Scientific Discovery* (1959). London and New York: Routledge Classics 2002.

Querenburg, B. v. 1973. *Mengentheoretische Topologie*. Berlin: Springer.

Quine, W.V. 1975. On Empirically Equivalent Systems of the World. *Erkenntnis* 9, 313–323.

Scheibe, E. 1982a. A Comparison of Two Recent Views on Theories. *Metamedicine* 3, 233–253 (slightly extended version of 1981c]. Repr. in: Scheibe 2001, 175–194.

Scheibe, E. 1987un. *Explanation, Reduction, Progress. Investigations into the Unity of Physics.* A Seminar given at the Philosophical Department of the University of California, Irvine 1987. Unpublished lecture. https://doi.org/10.17877/DE290R-22744.

Scheibe, E. 2001. *Between Rationalism and Empiricism. Selected Papers in the Philosophy of Physics*. Ed. by B. Falkenburg. Berlin: Springer.

Sneed, J. 1971. *The Logical Structure of Mathematical Physics*. Dordrecht: Reidel.

Stegmüller, W. 1976. *The Structure and Dynamics of Theories*. Berlin: Springer.

Chapter 4
Exact Reductions

The empirical reduction discussed in the previous section would perhaps be the only one a radical empiricist (or instrumentalist) would accept and consider necessary. For this reduction, and only it, makes use of the observation statements of the theories connected by a reduction. As has already been indicated, this is not sufficient from our point of view. Empirical reducibility will serve us as a kind of check that the theoretical reductions have a robust empirical basis. But we will not examine this empirical side of the matter even for all the kinds of reductions discussed in the following. It must suffice for us to do so for the most important cases, in order to show how the issue has to be understood in principle (cf. Sects. 4.1, 4.4, 5.1 and 5.3). This selective presentation has also to suffice in other respects. We should bear in mind that we already distinguished four main parts of a theory in Sect. 2.3, to which the previous chapter added the empirical description of a physical system as a fifth part. All these components participate in a reduction, so that for each kind of reduction one would have to trace the fate each of them suffers under the reduction. A further multiplier enters due to the particular feature of our procedure described in Sect. 1.3: We distinguish between as many kinds of elementary reduction as possible, which can then also be arbitrarily combined and, precisely in the important cases, indeed appear as combinations. It is impossible to treat the resulting variety here in anything like a comprehensive account. In this respect, we have to ask for general absolution from the outset, now that we are getting into the reduction business proper. The exact reductions to be examined in this chapter are those in which the reduction mechanism does not yet employ any approximations. Here, we proceed in considerable detail, given that it is important to introduce the subject proper of this study.

E. Scheibe, *The Reduction of Physical Theories*, Fundamental Theories of Physics 207,
https://doi.org/10.1007/978-3-662-65000-4_4

4.1 Direct Generalisations

The general idea of reduction by generalisation is to extend the range of possible physical systems considered up to a certain moment by incorporating *new* possibilities *and* seeing what gives the former their specificity in the context of the latter. Such an extension of a theory may concern its axioms, its core, or its domain of application. As it turns out, an extension of the domain is always in play if one wants to obtain a reduction. We speak of generalisations of the first, second, or third kind, depending on whether also the species of structure or the core (including the core axioms) is concerned, or only the domain of application. For the generalisations of the first kind, we restrict ourselves in this section to direct generalisations, in the sense that they do not yet include equivalences that affect the conceptual basis of a theory. The generalisations are throughout complemented by specialisations. Structurally simply the converse relationship, they however take place under different pragmatic circumstances: In the case of specialisations, one will first have the more general theory and then restrict it by a condition with which no 'insight' into the particularity of an already existing theory is associated.

Reduction by generalisation goes back to Aristotle (*Anal. Post. B*, Chaps. 1–10 and 13). The starting point is, for example, the concept of man and the question: What is man? For an explanation, the notion of man is first generalised, e.g., to the notion of a living being. Thus, we have the fact that

$$\text{man is a living being,} \tag{4.1}$$

and this is already an answer to the question of what man is. But animals are also living beings, and the crucial point is still missing. We now want to know what *distinguishes* humans among all living beings. An efficient distinguishing mark is rationality. For, not only do we have in parallel to (4.1)

$$\text{man is rational,} \tag{4.2}$$

but man is also the only rational living being, i.e.,

$$\text{a rational living being is human.} \tag{4.3}$$

Only by introducing the mark of rationality and the reversal (4.3) do we 'understand' in this context what a human being is. We reduce the concept of a human being to the (more general) concept of a living being (or: explain it by the latter) by figuring out what it is that makes a living being human: his or her rationality. A generalisation is, of course, already achieved by step (4.1), but a *reduction* through generalisation only by adding (4.2) and (4.3). If no misunderstandings are possible, however, we will also call the reductions by generalisation simply 'generalisations', for the sake of brevity.

The reductions of physical theories, which we will call *direct generalisations of the first kind*, follow the same pattern. They primarily concern the *axioms* (species of structure) of the theories. Let, first of all,

$$\Sigma \vdash \Sigma' , \tag{4.4}$$

so that Σ' admits all physical systems that Σ admits—but usually more. The distinction of the Σ cases among the Σ' cases is achieved by a formula γ (in the common arguments of Σ and Σ') so that, according to (4.2),

$$\Sigma \vdash \gamma , \tag{4.5}$$

and, corresponding to (4.3), finally, also, vice versa,

$$\Sigma' \wedge \gamma \vdash \Sigma . \tag{4.6}$$

Here, γ is in general assumed to be contingent with respect to Σ'. In case of consistent Σ, the implication $\Sigma' \vdash \neg\gamma$ is, anyway, excluded by (4.4) and (4.5). But, in the non-trivial cases, $\Sigma' \vdash \gamma$ is not admitted either, because only then is Σ' a genuine extension of Σ. Due to the inconvenience of excluding trivial cases, however, we want to admit this possibility, although only as a limiting case. The propositional form γ, which distinguishes Σ within Σ', is the *vehicle* of this kind of reduction (Sect. 1.3).

As a physical example, let us consider the generalisation of the velocity formula for the 1-dimensional free fall $x(t)$ of a mass point near the surface of the earth

$$\frac{1}{2}(dx/dt)^2 = g(1 + x_0/R)^{-1}(1 + x/R)^{-1}(x_0 - x) \quad (\text{for } \Sigma) \tag{4.7}$$

(with $0 \leq x \leq x_0$ and $x_0 = x(0)$), with g as the gravitational acceleration and R as the earth radius, by the acceleration formula:

$$(d^2x/dt^2) = -g(1 + x/R)^{-2} \quad (\text{for } \Sigma') . \tag{4.8}$$

(4.8) is more general than (4.7), in the sense that (4.8) follows from (4.7), as it has to be according to (4.4), and that, in (4.7), the initial value

$$dx/dt(0) = 0 \quad (\text{for } \gamma) \tag{4.9}$$

of the velocity is already fixed, as also follows and is required by (4.5). Conversely, (4.7) follows from the conjunction of (4.8) and (4.9). The proof is carried out with the energy law

$$\frac{1}{2}(dx/dt)^2 - gR^2(R + x)^{-1} = \frac{1}{2}(dx/dt)_0^2 - gR^2(R + x_0)^{-1}$$

resulting from (4.8). In reconstructing the theories belonging to (4.7) and (4.8), we consider g and R as constants and the motion $x(t)$ (and with it velocity and acceleration) as variable.

In Sect. 2.3, we saw that the original axiomatisation may not be formulated in the physical basic concepts proper, but only use auxiliary mathematical concepts. Although the set of conclusions formulated only in physical terms is thus 'in principle' determined, one can still be far from having an axiomatisation of these conclusions. And, even if one does have it, one would like to retain the mathematical axiomatics if it is particularly simple. These circumstances can make the production of reductions considerably more difficult, and we will, therefore, usually assume in general considerations that we have a genuine physical axiomatics. Already in the case of generalisations of the first kind, which concern the axioms, the problem becomes manifest, and here we will consider the case in which we have a genuine double axiomatics. Then, we have to take care that the following square closes:

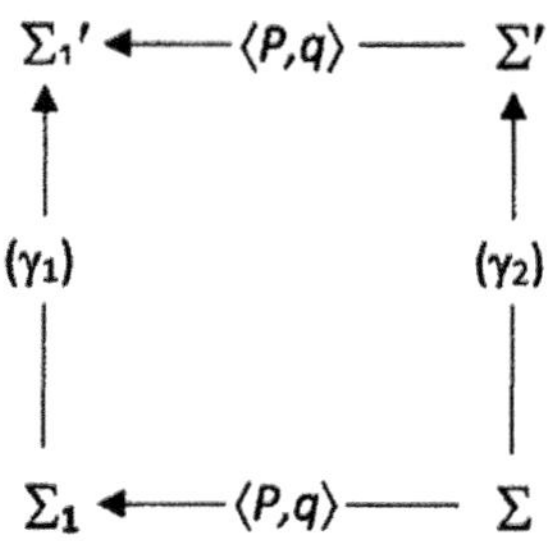

(4.10)

Here, Σ_1' is a generalisation of Σ_1 by γ_1 according to (4.4)–(4.6) and, correspondingly, Σ' of Σ by γ. Σ_1' and Σ' are the mathematical and physical axioms, respectively, of the reducing theory T', connected by the (conservative) representation $\langle P, q\rangle$ according to (2.57)–(2.62) of Sect. 2.3. Σ_1 and Σ are correspondingly related by $\langle P, q\rangle$, for the reduced theory T. Thus, we have direct generalisations, so to speak, on both levels, the mathematical and the physical, and in such a way that they *correspond* to each other. The correspondence is ensured above all by the fact that the mappings above and below in the square are *the same*. The 'closure' then means that the two paths in the square from the top left to the bottom right amount to the same thing.

Let us illustrate the square by electrodynamics. The mathematical axioms are formulated in terms of potentials, the physical axioms in terms of field intensities. On the other hand, electrodynamics is a unification of various theories that can be reduced to it, and these reductions also contain generalisations of the first kind (Sect. 4.5). At the moment, we understand electrodynamics only as a generalisation of electrostatics and (the latter) still in the version containing the magnetic field and the current. Since a static field specifies an inertial system, we write the potentials, field intensities, and current densities in the split form $\langle \varphi, \mathbf{A}\rangle$, $\langle \mathbf{E}, \mathbf{H}\rangle$, and $\langle c\rho, \mathbf{j}\rangle$, respectively. Then, in the potentials, we have the equations ($\partial\cdot$ for $\partial\cdot/\partial t$)

$$
\begin{aligned}
&-c^{-2}\cdot\partial^2\varphi+\nabla^2\varphi=-4\pi\rho\\
&-c^{-2}\cdot\partial^2\mathbf{A}+\nabla^2\mathbf{A}=-4\pi\mathbf{j}\\
&c^{-1}\cdot\partial\varphi+\mathrm{div}\mathbf{A}=0\quad\text{(Lorentz convention)}
\end{aligned}
\tag{4.11}
$$

as mathematical axioms (Σ'). Using the representation

$$
\mathbf{E}=-\mathrm{grad}\varphi-c^{-1}\cdot\partial\mathbf{A},\quad \mathbf{H}=\mathrm{rot}\mathbf{A}
\tag{4.12}
$$

(for $\langle P, q\rangle$), we then obtain Maxwell's equations

$$
\begin{aligned}
&\mathrm{rot}\mathbf{E}=-c^{-1}\cdot\partial\mathbf{H},\quad \mathrm{div}\mathbf{H}=0\\
&\mathrm{rot}\mathbf{H}=c^{-1}\cdot\partial\mathbf{E}+4\pi c^{-1}\mathbf{j},\quad \mathrm{div}\mathbf{E}=4\pi\rho
\end{aligned}
\tag{4.13}
$$

as physical axioms Σ'. The specialisation

$$
\partial\mathbf{E}=\mathbf{H}=\mathbf{j}=0
\tag{4.14}
$$

(as γ) then gives the electrostatic equations

$$
\mathrm{rot}\mathbf{E}=0,\quad \mathrm{div}\mathbf{E}=4\pi\rho\,,
\tag{4.15}
$$

where Σ is given by these equations *and* (4.14) in order to satisfy (4.5). On the other hand, we can first specialise (4.11) by

$$
\partial\varphi=\mathbf{A}=\mathbf{j}=0
\tag{4.16}
$$

(as γ_1) to

$$
\nabla^2\varphi=-4\pi\rho\,,
\tag{4.17}
$$

where again the last two lines make up Σ_1. The representation (4.12) now also carries (4.16) and (4.17) over into (4.14) and (4.15).

As already mentioned at the beginning, the *core* and also the *core axioms* of a theory are preserved in the case of generalisations of the first kind. So, here, we have, in the notation of (2.47)–(2.49) of Sect. 2.3,

$$
X^{\circ\prime}=X^{\circ},\ s^{\circ\prime}=s^{\circ},\ \Sigma^{\circ\prime}=\Sigma^{\circ}.
\tag{4.18}
$$

With the domain of application $\mathfrak{N}$, things are different again. Whereas the equalities (4.18) agree with the reduction by generalisation, although they do not affect the core, we would in general not arrive at a reduction regarding the domain of application with the equality $\mathfrak{N}=\mathfrak{N}'$. For if $\mathfrak{N}'$ is at the same time the scope of T' (as is aimed at and is possible in principle), then with a genuine generalisation $\mathfrak{N}$ will just not be equal to $\mathfrak{N}'$. The assertion of the theory T for $\mathfrak{N}$ would be false for certain systems, since γ is not satisfied, and T could not be reduced to T' with this domain of application.

Rather, by generalising T to T', one learns just to which cases T is to be restricted, namely, those that satisfy γ. The correct relationship between $\mathfrak{N}$ and $\mathfrak{N}'$ for the reduction is, therefore,

$$\mathfrak{N} \subseteq \mathfrak{N}'/\gamma\,, \tag{4.19}$$

where $\mathfrak{N}'/\gamma$ are those systems of $\mathfrak{N}'$ that satisfy γ. However, one does not have this relation before knowing the generalised theory. But this is also not to be expected: It is generally impossible to know the scope of a theory before having a better theory. Due to (4.6), from (4.19), it now follows, for the theoretical assertions (2.69) of Sect. 2.3, that

$$\mathrm{Th}'(\mathfrak{N}') \vdash \mathrm{Th}(\mathfrak{N})\,. \tag{4.20}$$

In a certain context, the following procedure is easily confused with a direct generalisation (4.4)–(4.6), (4.18) and (4.19), but, in fact, it is a different kind of reduction. If, for example, we are dealing with the theory of an ideal gas, i.e., essentially with the equation $pv = RT$ with constant R and variables p, v, T, we can, for example, convert the variable T into a constant T_0 and draw the conclusion

$$pv = RT \wedge T = T_0 \vdash pv = RT_0\,.$$

This conclusion has the form (4.6). But we would not also have (4.4) and (4.5) (in this form). And, above all, our two theories would not have the same constants and variables. For we have passed over from the ideal gas equation (almost) to Boyle-Mariotte's law (with constant temperature). So we are already dealing here with generalisations of a different kind, in which constants are replaced by variables while *retaining* the axioms.

In general, a *generalisation of the second kind* is about leaving the species of structure Σ unchanged *in a certain sense*, replacing the axiom

$$\Sigma(X^\circ; s_1^\circ \ldots s_m^\circ, s_1 \ldots s_n)$$

of theory T by the axiom

$$\Sigma(X^\circ; s^\circ \ldots s_{m-1}^\circ, s_0 \ldots s_n)$$

of theory T', i.e., the constant s_m° is replaced by the variable s°, according to

$$\Sigma' \equiv \Sigma(\ldots s_0 \text{ for } s_m^\circ \ldots) \tag{4.21}$$

and *insofar* as Σ also changes. This change is produced by the fragmentation

$$X^{\circ\prime} = X^\circ,\; s^{\circ\prime} = s^\circ \text{ without } s_m^\circ \tag{4.22}$$

of the core of T. If we now require, in addition,

$$\mathfrak{N} \subseteq \mathfrak{N}'(s_0^y = s_m^\circ) \,, \tag{4.23}$$

according to (4.19), we again obtain the implication (4.20)

$$\mathrm{Th}'(\mathfrak{N}') \vdash \mathrm{Th}(\mathfrak{N})$$

that expresses the validity relation between both theories.

Finally, let us consider, as *generalisations of the third kind*, those theory relations in which

$$\Sigma' \equiv \Sigma, \quad X^{\circ\prime} = X^\circ, \quad s^{\circ\prime} = s^\circ, \quad \Sigma^{\circ\prime} \equiv \Sigma^\circ \tag{4.24}$$

holds and only the domain of application is extended:

$$\mathfrak{N} \subseteq \mathfrak{N}' \,. \tag{4.25}$$

Again, as above, then immediately (4.20)

$$\mathrm{Th}'(\mathfrak{N}') \subseteq \mathrm{Th}(\mathfrak{N})$$

results for the assertions of the theories. The mere extension of the domain of application of a theory is, in itself, of no theoretical interest. However, it is clear that it plays a significant role in the practice of research since we are interested in pushing the limits of validity of the theories as far as possible. Again and again, to draw conclusions, we presuppose that the physically real possibilities are also formal possibilities in the sense of the axioms of the theory (and, vice versa, that the formal possibilities are also real). Under certain circumstances, this closure can only be achieved by a generalisation (4.25).

We will now go on to consider, above all, the direct generalisations of the first kind. Almost everything that needs to be said about them also applies to the indirect generalisations of the first kind dealt with in the following sections. But we say it already here because we have other concerns later. First of all, we have to correct the popular philosophical opinion that the basic formal condition for a reduction is that the reduced theory follows from the reducing theory (cf. Sect. 1.2). That something is wrong with this view can already be seen from the generalisations of the first kind, even though here we are still remaining in the domain of exact reductions and do not yet have the argument of the incompatibility of reduced and reducing theory available. In our reconstruction, conclusion (4.20) about the assertions of the theories T and T' comes closest to that popular opinion. But, obviously, it is not the case that the *axioms* of the reduced theory would follow from those of the reducing theory. For the axioms, the *reverse* is true because of (4.4). It is only together with the additional condition γ that Σ follows from Σ'. In fact, only (4.4) is used for the proof of (4.20), but even this relation is much too weak to justify a reduction on its own. In general, (4.6) does not cover all models of Σ and, thus, not all physical systems admitted by Σ. Therefore, one must always take into account that among them there are empirical instances that cannot be explained by any model of $\Sigma' \wedge \gamma$. Thus, once one starts with (4.6), one must also accept (4.4) in order to achieve a reasonable reduction, and, as one can see, this almost results in a reduction by generalisation as is then achieved with (4.3). From this point of view, (4.20) is an important conclusion, which is significant for the progress associated with a reduction (see below), but which depends on the underlying, truly fundamental relations. In view of the fact that (4.20) also occurs in the case of generalisations of the second and third kind, we can suppose it, if correctly formulated, to be an *adequacy condition* for exact reductions.

Despite these differences between the reductions by generalisations of the first kind and DN-explanations (cf. Sect. 1.2), generalisations of the first kind may have a considerable *explanatory value*. Because of the formal similarity of condition (4.6) with singular explanations in the sense of Sect. 3.1, one might think that the explanatory value in question is the same here as there. This is the position Hempel and Oppenheim took in 1948 in their conception of the DN explanation, which at that time and since then has been explained primarily based on singular explanations, but was in principle not limited to them (Hempel 1965, 247–248 and 273). In our approach, however, further differences are striking. It is obvious that a singular explanation does not generalise and that a generalisation does not involve singular statements. The explanation of a singular statement is primarily achieved by the additional premise of the deduction scheme, which is also singular in many cases, with the help of the explanatory theory. In contrast, our theory reduction is primarily carried out by the generalised theory, with the help of the respective additional premise. This different attribution of one of the two premises to the respective explanandum becomes particularly clear in elliptical formulations of the explanations in question. We explain the broken window pane by the thrown stone—one contingent event by another—and usually do not mention at all the theory by means of which the explanation is made. In contrast, we speak of the explanation of Kepler's laws by Newton's theory—one generality by another—, and *here* we do *not* mention the specific conditions that only allow for the explanation.

Nevertheless, the structural similarity between singular and generalising explanations is important for the insight that both cases indeed concern explanations. Just as in the former context, *many* singular facts can exist within the scope of *one* theory, and we get an answer to the question "why this one and not that one?" through the initial and boundary conditions that, together with the theory, make that fact necessary, so here we understand the why of a theory Σ in competition with others (with the same generalisation Σ') likewise, through the condition γ that is decisive within the scope of Σ' for Σ to occur. And in both cases—the singular explanation and the present one—this kind of understanding takes place against the background that possibilities other than those distinguished by the respective condition actually occur. As we shall see, however, the explanatory power of generalisations depends on the premise γ in relation to Σ and Σ' and, thus, on pragmatic conditions. In contrast to singular explanations, γ is not already clearly distinguished from the theory axioms by its own status. In limiting cases, for example, it can be *equal* to Σ, in which case one would certainly not speak of an explanation.

The conditions (4.4)–(4.6) are in themselves *symmetric* in Σ' and γ. However, we have the feeling that the *asymmetry* prevailing in singular explanations, and so easily conceived there, also exists in our case of generalisations. The problem of what it consists of is not precisely solved here. In the evolution of physics and, in fact, of science in general, the procedure of generalisation is in principle significant because it introduces to us hitherto unknown and, therefore, unconsidered possibilities—unconceived alternatives to a hitherto prevailing theory. This significance is only in principle because, of course, not everything which is conceivable is also capable of grasping the actual world. But, if something conceivable actually is—in a given case—, then one has taken an important step forward. The most important obstacle to this kind of progress comes from attempts at a so-called justification of theories. For such a justification has the aim of showing that there are no alternatives. Kepler's attempt to explain the distances of the planets from the sun was intended to prevent the idea that the distances do not matter, that by theory they could differ from what they are. The many attempts to present Euclidean geometry as necessary in some sense, culminating in Kant's so-called a priori justification (and more recent ones), did certainly not favour the emergence or use of non-Euclidean geometries. It may be that, one day, we will no longer need to generalise because any possible experience has been made in one field of knowledge. Then, the business of justification could begin here not only without harm but also with benefit (Weizsäcker 1980, Part II). But probably it is, in principle, impossible to know when this moment will have come.

Another analogy with singular explanations is of great importance. For the latter, it is a much discussed topic that they are structurally equal to *predictions* (Hempel 1965, 364–376). The difference between explaining and predicting an event seems to be essentially pragmatic. If we understand explanations and predictions as occurrences in the scientific world, an explanation requires that the event to be explained is *given* at the time of the explanation. Prediction, on the other hand, is what its name implies, precisely if the event to be predicted is still in the *future* at the time of the prediction. Nevertheless, the logical relations are the same in both cases. In the case of theory explanations and reductions, we have a corresponding situation. The usual understanding of the words 'explanation' and 'reduction' presupposes here, too, that the explaining or reducing theory T already *exists* for the purpose of this manipulation, while the theory T' to be explained or reduced is still *to be found* along with further ingredients (vehicles, etc.). Perhaps even more frequent than this procedure is the converse one in which one obtains a theory T from a given theory T' according to the same kind of operation that would have been applied in order to reduce the initially given theory T to the theory T'. For the reductions that we call 'generalisations', the converse reduction is pretty much what we call *specialisation*. Generalisation and specialisation are logically correlative terms, so that whenever T' is a generalisation of T, we can also express the same (logical) fact by saying that T is a specialisation of T'. But, if we understand generalisation and specialisation as procedures in scientific life, we have at least the pragmatic difference that in the first case T is given and T' is sought, in the latter vice versa. In accordance with the historical significance of generalisations, specialisations can also become important if, following a generalisation of theory T to theory T, they lead to further, hitherto unknown theories $T_1, T_2, \ldots$ as special cases of T'. Let us think, for example, of general classical mechanics and quantum mechanics. Furthermore, specialisations occurring together with approximations, which lead in particular to mathematical simplification, are invaluable for the progress of physics; cf. Chap. 5.

The pragmatic difference between generalisation and specialisation as operations has consequences with regard to the degree of insight and explanatory power associated with the equivalence

$$\Sigma' \wedge \gamma \equiv \Sigma \,, \tag{4.26}$$

to which (4.4)–(4.6) amounts. As we have seen, specialisations are obtained by tentatively choosing the additional condition γ (with the intended application in mind), starting from Σ', in order to *identify* Σ simply with the conjunction $\Sigma' \wedge \gamma$. This is obviously an extremely trivial case of (4.26), in which, moreover, Σ cannot be known prior to Σ'. If, on the other hand, we are dealing with the generalisation of a given Σ, then Σ was historically known prior to Σ' and its formulation can, thus, not have made use of Σ'. For generalisations, one will, therefore, expect to come across non-trivial cases of (4.26) as explanations (as far as such simple relations as given by (4.26) may have at all explanatory power). We will get to know cases of this kind in the next section. In any case, we see how the fact that (syntactic) conditions of the components of (4.26) come into play here results in a 'qualitative graduation' of the insight provided by (4.26). Although this graduation cannot be made precise, it suffices to show how futile the dichotomies of 'genuine' explanations and non-explanations are. In particular, it would be of little use to exclude altogether the limiting case in which (4.26) becomes a (literal) identity.

By the way, we obtain a similar graduation of the explanatory power by considering the relation between Σ, Σ', and the condition γ, in (4.26). In a generalisation (as a procedure), Σ and Σ' are known prior to γ. Equation (4.26) can then be understood as an equation for the unknown γ under the constraint (4.5). Here, we have (among others) the trivial solution Σ, which has zero explanatory power, just as the case of a plain identity in (4.26). For, according to an earlier consideration, we want to explain Σ by getting to know within the scope of Σ' the special condition that restricts Σ' to Σ. And it is certainly useless to be given Σ as this condition. On the other hand, as far as the relation between γ and Σ' is concerned, the latter is known prior to γ both for generalisations and for specialisations. Therefore, it makes sense to require that γ be logically independent of Σ'. Apart from the already discussed case that γ is contingent relative to Σ', the possibility $\gamma \vdash \Sigma'$ still has to be considered here. Since it results in $\gamma \equiv \Sigma$, one will want to exclude it. Again, however, we can already see at this point that this, too, is a matter of more or less, which therefore should not be fixed.

It is also not the case that the insight provided by a generalisation must inevitably be due to finally finding γ from (4.5), with Σ given and after finding first Σ' (i.e. without γ coming in between), such that the 'surprise' lies in the validity of (4.6) with this γ. The step from Euclidean to non-Euclidean geometry, as it turned out to be for the mathematicians of the 19th century, was certainly a step of generalisation.

Here, however, Σ as Euclidean geometry and γ as the parallel axiom had been known for a long time, and at least in retrospect the new insight was to find a representation $\Sigma = \Sigma' \wedge \gamma$, combined with the thought that the new combination $\Sigma' \wedge \neg\gamma$ was at least a logical possibility. In particular, a perfect formulation of γ here preceded that of Σ', the absolute geometry. Once having the latter—but only then—, the step from Euclidean to absolute geometry, and more generally the mere 'omission' of γ in an already known axiomatic $\Sigma' \wedge \gamma$, offers itself as a particularly simple example of direct generalisation.

Finally, we have to turn to the question of *empirical progress* regarding generalisations. According to the general preliminary considerations in Sect. 3.3, we have to show that the three kinds of generalisations so far discussed make empirical progress *possible*. This means, first and foremost, that possible single successes (by measurement results) of the reduced theory T are also successes of the reducing theory T', and that even compatibilities of T (with measurement results) remain such with respect to T'. Since we are still dealing with exact reductions here, we can also refer to exact successes and confirmations (in the sense of Sect. 3.1) in the following. However, the situation would remain the same if we used approximate successes and confirmations instead. Moreover, as is immediately obvious, what kind of concept of success we invoke at all is largely irrelevant for generalisations of the second and third kind. The preservation of possible empirical individual successes is based here solely on the fact that the domain of application of T is a subset of the domain of application of T' according to (4.23) and (4.25), whereas the theoretical axioms for the single system are essentially the same. Initially, one could think that tests of, e.g., Boyle-Mariotte's law of gases are not also tests of the full law of an ideal gas and, thus, of a generalisation of the second kind, given that the former only includes pressure and volume, whereas for an ideal gas the temperature must also be included. But, once we have realised that the former law is a special case of the latter *for a given temperature*, then we can take a test of the former as a test *at this temperature* and thus place it among the tests for the full gas law. In a sense, the temperature has already been decided. Because of (4.21), successes of the more specific law are also trivially successes of the more general law.

In this respect, there is a problem only for the generalisations of the first kind. For in this case, Σ' makes a weaker statement about the single system than Σ because of (4.4), and this can result in the fact that via Σ' a test (3.5) does not establish a dependence (3.12) (of Sect. 3.1), even if Σ does. As an example, we choose geometry. In Euclidean geometry, the theorem holds that the sum of angles in a triangle has the value π (in radians). If one admits the angles as derived quantities for measurement and for checking the geometry, three measured angles α, β, γ would be a success if

$$\alpha + \beta + \gamma = \pi \tag{4.27}$$

holds. After all, one could calculate and predict each of the three angles from the other two. This possibility disappears when we move from Euclidean geometry to absolute geometry, which is a direct generalisation that omits the parallel axiom (cf. Borsuk and Szmielew 1960). We would thus have a case in which possible successes of a theory would not be possible successes of a direct generalisation. From this failure of the conservation principle for successes, one may draw the far-reaching consequence of restricting the notion of generalisation in such a way that the principle remains valid. But to take this step, one must already be very confident of one's concept of corroboration. For above all, it must be expected that the same situation will also arise for completely different kinds of reduction so that one would consequently have to restrict these as well, and the whole story would have incalculable consequences for the concept of reduction.

In this respect, it is better to leave things as they are and to take another look at the concept of corroboration—and, by changing it, to enforce the principle of conservation in question. Strictly speaking, it is not even a question of changing the concept of corroboration itself, but only its domain of application. In Sects. 3.1 and 3.2, we deliberately did not address the problem that now arises because its main reason is at the same time our main topic, namely reduction, and, therefore, an anticipation of it as a whole would have been inappropriate. However, we now advocate for an extension of the concept of success such that successes for a theory T' are also always given if *direct* successes, in the sense of 3.1, are given for a *special case* T of T', or: that these successes of T are also considered *indirect* successes of T'. The empirical validity of (4.27) would, therefore, also be a success of absolute geometry, not as a direct success in the sense of Sect. 3.1 but an indirect one via the special case of Euclidean geometry. (As a success of absolute geometry, however, it would have to be flanked by successes of the other, non-Euclidean geometries).

This would save our conservation principle, even if this procedure looks rather ad hoc. But this is due more to the fact that our main problem here is not the concept of empirical corroboration but that of reduction. If it were mainly about the former, we would undoubtedly widen our view and take into account, in particular, the principle that successes of special cases of a theory are also always successes of the theory itself. Every physicist is likely to think so, and we are not talking about anything else here. Of course, we do not believe that arbitrary *generalisations* of a given T are physically meaningful and that the concept of success should, therefore, also be extended into the unmeasured. But, once a theory T' is regarded as physically possible, then this is also transferred to its *special cases*, and this fact justifies the disputed extension of the concept of success.

The conservation principles only serve to prevent any deterioration in the direction of the reduction (cf. Sect. 3.3). What about the possibility of genuine improvements of generalisations? For generalisations of the second and third kind, obviously the failures are preserved, as well as the successes, and an improvement of T is only possible through new successes in the larger domain of application $\mathfrak{A}'$ of T'. The situation is different with generalisations of the first kind. Here, too, the improvement can consist in achieving new successes in a larger domain of application. But it can also consist in transforming failures of T into successes of T'. Newton's generalisation of Kepler's laws provides this possibility. Kepler's theory can be defined in modern form by the conditions

$$\begin{aligned} d^2\mathbf{x}/dt^2 &= -\mu\mathbf{x}|\mathbf{x}|^{-3} \\ \frac{1}{2}|d\mathbf{x}/dt|^2 - \mu|\mathbf{x}|^{-1} &< 0 \end{aligned} \tag{4.28}$$

for the motion $\mathbf{x}$ of the planet. Concerning it, by Newton's generalisation, we only mean the theory that arises from (4.28) by omitting the second line for the moment. This adds parabolic and hyperbolic orbits, and just they make observation statements about them possible which are so strong that they contradict Kepler's theory. But they are so strong precisely if they are at the same time a success of Newton's theory. In general, it is here the weakening of Σ to Σ' that makes this transformation of failures into successes possible. If we require (4.19), however, the failures thus eliminated cannot lie in the domain of application $\mathfrak{R}$ of T. This relation already presupposes the 'purification' of $\mathfrak{R}'$, and the failures in question do indeed not satisfy γ.

4.2 Equivalences

Measured against the usual way of thinking and speaking, as well as against a multitude of cases actually occurring in physics, the version (4.4)–(4.6) of the generalisation of the first kind and the corresponding reduction is too narrow. Let us illustrate this by means of the case that in the history of physics has probably been the most significant relevant case. Later, it will turn out, anyway, that the strict deductive reduction by generalisation (including the extension to be made now) is not sufficient to cover precisely the historically significant cases of progress and unification by reduction. But *one* such case undoubtedly belongs here, and that is the generalisation of Euclidean to Riemann geometry, as well as the later parallel case of the generalisation of Minkowski's to Lorentz's geometry, i.e., to the theory of Lorentz manifolds. The fundamental idea, already anticipated by Gauss in the theory of surfaces and then formulated in a very general and abstract way by Riemann, was to cease requesting Euclidean relations for the whole of space nor only for finite areas, but only "in the infinitely small", i.e., no longer strictly at all, but only in the limiting case of infinitely small extents, which may never be realised exactly. It is clear that the concept of Riemannian space conceived according to this idea had to contain elements that were not only alien to the original conceptual world of Euclidean geometry, but were neither obvious in geometric use even after introducing the infinitesimal calculus.

Thus, if today we regard Riemann geometry without further ado as a generalisation of Euclidean geometry, then this cannot be meant in the sense of the definition (4.4)–(4.6) that presupposes complete conceptual homogeneity for Σ and Σ': A Euclidean space in the pre-Riemannian sense, even if already understood as structure in a modem sense, is certainly another kind of structure than a Euclidean space in the sense of a *direct* specialisation (4.4) of a Riemann space. But given a sufficiently strong 'logic'—in our case set theory—, (old) Euclidean geometry turns out to be *equivalent* to a theory that is a direct case (in the sense of the previous section) of Riemann geometry. With Σ as Euclidean geometry, Σ' as the Riemannian version of it, and Σ'' as Riemann geometry in general, we have Σ'' as an *indirect generalisation* of Σ:

$$\begin{array}{ccc} \Sigma & \xrightarrow{\textbf{(equiv)}} & \Sigma' \\ & & \Big\downarrow \textbf{(dir.gen.)} \\ & & \Sigma' \end{array} \tag{4.29}$$

This, by the way, would be a first *combination* of reductions, if we also want to regard equivalence as a reduction.

In this section, we will indeed deal almost exclusively with the concept of equivalence and only in the next section come to the indirect generalisations. It is clear from the outset that equivalences are reductions in a degenerate sense at best. For each of two equivalent theories can be reduced one to the other—or so it seems, and thus the usual asymmetry of the reduction relation would be missing here. Thus, it is important to first clarify by means of a clear reduction, such as the generalisation, which role equivalences may play in the business of reduction. They play a role mainly in combinations, not only with direct generalisations but also with other reductions proper. This role then lies in the extremely important procedure of *conceptual adaptation* and uniformisation—in our example, the adaptation of the Euclidean conceptual world to the Riemannian one, as well as a uniformisation of both for the special case of Euclidean geometry. Actually, the equivalence (4.4)–(4.6) of $\Sigma' \wedge \gamma$ with Σ already arises in the direct generalisations of the previous section. But, for this equivalence, we had assumed that its two sides are built up from the same conceptual material, and thus a conceptual adjustment is no longer required. Only in (4.29), with the step from Σ to Σ', is such a procedure envisaged concerning Σ''.

Let us begin with an example in which the axioms of the theory are expressed in terms of defined concepts and in which equivalences occur because different defined concepts are used for the same basic conceptual material. The example is Kepler's theory of planetary motion and its 'Newtonian' reformulation. In both cases, the physical system is described by two parameters: the motion $\mathbf{x}$ of the planet and the Kepler constant μ. Using defined terms, Kepler's laws proper require for $\mathbf{x}$ and μ:

Kep I: $\mathbf{x}$ is (spatially) an ellipse with the zero-point as one focal point.
Kep II: $\mathbf{x}$ has (temporally) constant surface velocity.
Kep III: For the major semi-axis a of the ellipse and the orbital period T,

$$a^3/T^2 = \mu/4\pi^2.$$

These laws now permit an equivalent formulation that no longer employs the geometric concepts of ellipse, surface velocity, etc., but rather the more recent Newtonian concepts of orbital velocity and acceleration. In terms of them, the requirements for $\mathbf{x}$ and μ look as follows. We have the equation of motion

$$d^2\mathbf{x}/dt^2 = -\mu\mathbf{x}|\mathbf{x}|^{-3} \tag{4.30}$$

and the inequality

$$\frac{1}{2}|d\mathbf{x}/dt|^2 - \mu|\mathbf{x}|^{-1} < 0, \tag{4.31}$$

which restricts $\mathbf{x}$ to periodic motions. The equivalence claim is here ultimately, of course, that the laws Kep I-III have the *same* solutions $\langle \mathbf{x}, \mu \rangle$ as (4.30) and (4.31) (cf. Born 1949, Appendix 1). Along with this equivalence, however, there is also an indirect generalisation (4.29) of Kepler's laws to Newton's equation (4.30), which has the inequality (4.31) as explanatory premise γ.

For, due to the occurrence of *different* defined concepts in the respective axioms, we do not have the equivalence already tacitly used in the previous section, that is, equivalence as mere a re-axiomatisation in terms of the *same* concepts. What we have now may be expressed in general as follows:

$$\begin{aligned} s_1 = r(X; s) \wedge \Sigma(X; s, s_1) \wedge s_2 = q(X; s) \vdash \Sigma'(X; s, s_2) \\ s_2 = q(X; s) \wedge \Sigma'(X; s, s_2) \wedge s_1 = r(X; s) \vdash \Sigma(X; s, s_1) \end{aligned} \tag{4.32}$$

Here, in the terms r and q, the equations provide the respective definitions in Σ and Σ', respectively. An equivalence (4.32) concerns the exchange of concepts defined in terms of basic concepts, where the latter remain fixed. This is already a case of mutually adapting Σ and Σ'. What is really interesting, however, is the case where the basic concepts themselves, but only the typified ones, are exchanged. Let us first consider two examples.

In the theory of black body radiation, *Planck's radiation law* is given as a relation that connects the temperature of the radiation with its spectral energy density (cf. Schpolski 1954, Sect. VI). The spectral energy density indicates how the energy is distributed among the components of the radiation. The components can now be described either by their frequencies or their wavelengths. The energy density in question is, thus, either a function of the frequency or the wavelength. Depending on which representation we choose, we arrive at one or the other formulation of the law of radiation, which are, of course, physically equivalent. For the temperature T and the energy density ρ as a function of the frequency ν, we obtain

$$\rho(\nu) = c^{-3}8\pi\nu^3\{\exp(h\nu/kT) - 1\}^{-1}\,. \tag{4.33}$$

With ρ' as a function of the wave length ν, on the other hand, we obtain

$$\rho'(\lambda) = 8\pi c\lambda^{-5}\{\exp(hc/kT\lambda) - 1\}^{-1}\,. \tag{4.34}$$

The relation between them is given by the fact that, in each case, $\rho(\nu)d\nu$ or $\rho'(\lambda)d\lambda$ represents the fraction of the total energy falling into the interval $d\nu$ or $d\lambda$. For $\lambda\nu = c$, these fractions must be equal:

$$\rho(\nu)d\nu = \rho'(\lambda)d\lambda \ \ (\text{for } \lambda\nu = c) \ . \tag{4.35}$$

This gives the transformation formulae

$$\begin{aligned} \rho'(\lambda) &= \rho(c\lambda^{-1})(c\lambda^{-2}) \ , \\ \rho(\nu) &= \rho'(c\nu^{-1})(c\nu^{-2}) \ , \end{aligned} \tag{4.36}$$

from which the equivalence of (4.33) and (4.34) follows. If we denote (4.33) and (4.34) by $\Sigma(\rho, T)$ and $\Sigma'(\rho', T)$, respectively, and the transformations in (4.36) by $q(\rho)$ and $q^{-1}(\rho')$, respectively, then the whole issue of equivalence is captured by the formulas:

$$\begin{aligned} \Sigma(\rho, T) \wedge \rho' = q(\rho) &\vdash \Sigma'(\rho', T) \ , \\ \Sigma'(\rho', T) \wedge \rho = q^{-1}(\rho') &\vdash \Sigma(\rho, T) \ . \end{aligned} \tag{4.37}$$

(Due to the symmetry in $\nu\lambda = c$ here, in particular, $q^{-1} = q$ holds.) To clarify the background of the two theories a bit more, we note that the base sets forming the core are (in both cases) the temperature scale and, in addition, for (4.33) the frequency scale and the scale of energy density per frequency, but for (4.34) the scales of wavelength and energy density per wavelength. To make the transition from one theory to the other possible, however, these lists (per theory) would be too scarce. Without using a deeper theory, we could not separate the wavelength scale and the frequency scale, nor the energy density scales. Together with the relation $\rho\nu = c$, we must rather add the respective other base sets to each of the two theories, although only one pair is sufficient to express the radiation law. We will often encounter such situations of conceptual uniformisation and permit us corresponding liberties.

A widely known and already slightly more complicated example of a physical equivalence is that between *Hamiltonian* and *Lagrangian mechanics*, both restricted to conservative and time-independent forces. The time and the configuration space M (and thus, as it has to be, the principal base sets) remain fixed in the equivalence transformation. The rest takes place on two different stages: for Hamiltonian mechanics on the cotangent bundle T^*M, for Lagrangian mechanics on the tangent bundle TM. Thus, according to Hamilton, we have the kinetic energy T as a Riemann metric

$$T : T^*M \to \mathbb{R}, \ T = \frac{1}{2}\sum g^{ik} p_i p_k \tag{4.38}$$

and, according to Lagrange, as a Riemann metric

$$T : TM \to \mathbb{R}, \ T' = \frac{1}{2}\sum g_{ik} v^i v^k \ . \tag{4.39}$$

Equivalence holds (insofar as) if we are dealing with *the same* metric, i.e., if the equivalence transformations

$$v^i = \sum g^{ik} p_k, \quad p_i = \sum g_{ik} v^k \quad (g^{ik} g_{kl} = \delta^i_l) \tag{4.40}$$

are inverses of each other. Furthermore, we have the Hamiltonian

$$H : T^*M \to \mathbb{R} \text{ with } \partial H/\partial p_i = v^i\ , \tag{4.41}$$

on the one hand, and the Lagrangian

$$L : TM \to \mathbb{R} \text{ with } \partial L/\partial v^i = p_i \tag{4.42}$$

on the other. Here, we have given the additional conditions in 'mixed' notation, as usual. What is meant in each case is the equation that results from substituting (3.48) for p_i or v^i. The conditions are equivalent to $H - T$ and $T' - L$ being independent of p_i and v^i, respectively. The connection between H and L is that $H - T$ and $T' - L$ concern *the same* functions, namely, the potential energy of the system, which depends only on the coordinates in M. This state of affairs is usually written in the form

$$H = \sum p_i v^i - L\ , \tag{4.43}$$

where, again, the equivalence transformations (4.40) are to be taken into account: If read from right to left (introduction of H), v^i must be replaced by its definiens, but if resolved according to L, the same applies to p_i. Finally, the curves

$$C : \mathbb{R} \to T^*M \text{ with } dq^i/dt = \partial H/\partial p_i,\ \ dp_i/dt = -\partial H/\partial q^i \tag{4.44}$$

and

$$C' : \mathbb{R} \to TM \text{ with } d(\partial L/\partial v^i)/dt - \partial L/\partial q^i = 0 \tag{4.45}$$

have to be considered. These are also transformed into each other by the equivalence transformations (4.40), but it is important to ensure that the projections of C and C' in M are *the same* curves.

If we now consider the *general case*, the equivalence of species of structure will essentially involve transformations of the form

$$\begin{aligned} \Sigma(X, s) \wedge t = q(X; s) \vdash \Sigma'(X; t)\ , \\ \Sigma'(X, t) \wedge s = q^{-1}(X; t) \vdash \Sigma(X; s)\ , \end{aligned} \tag{4.46}$$

where q and q^{-1} for Σ and Σ', respectively, are internal terms in the sense of (2.57)–(2.59) of Sect. 2.3, for which, moreover, the inversion conditions

$$\begin{aligned} \Sigma'(X;t) &\vdash q(X;q^{-1}(X,t)) = t \\ \Sigma(X;s) &\vdash q^{-1}(X;q(X;s)) = s \end{aligned} \tag{4.47}$$

hold. It is crucial that the principal base sets do not change and that in this framework all structures of the kind Σ are uniquely correlated with all structures of the kind Σ'. The core $\langle X^\circ; s^\circ \rangle$ of the theory and the corresponding species of structure Σ° can be affected by an equivalence transformation, but do not have to be. Our example (4.33)–(4.37) of two versions of Planck's law is certainly of the latter kind. On the other hand, we can conceive of a Hamiltonian and a Lagrangian in such a special way that the Hamiltonian and the Lagrangian belong to the core and only the trajectories are variable. Then, with (4.38)–(4.45), we have an equivalence in which the core changes.

The most important question is, of course, by what right we classify equivalences as reductions. First of all, it seems that only *one* of the two lines of (4.46) can stand for *one* reduction so that, at least verbally, a reduction direction exists: The first line would reduce Σ' to Σ, the second Σ to Σ', where q or q^{-1} would be the respective vehicle of the reduction. If at all—so it seems—we would be dealing here with a *reciprocal* reduction. But, precisely on this point, there are reasons to be cautious. First, of course, we cannot base our justification solely on the fact that each of the two lines of (4.46) is already a (set-theoretic) *deduction*. For, in the previous section, we have become acquainted with substantial reasons against equating deduction and reduction, *even if* only the formal aspects of a reduction are concerned. To each of the two deductions in (4.46), however, is added the fact that there is the other, and this has consequences which bring us closer to an interpretation as a reduction.

Thus, from the second line of (4.47), follows

$$\Sigma(X;s) = \Sigma(X;s') \wedge q(X;s) = q(X;s') \vdash s = s' \tag{4.48}$$

and, hence, that q is *injective* with respect to Σ. If now (4.48) is added to the deduction in the first line of (4.46), then formally we have before us the operation of *embedding* Σ in Σ'. In the next section, we will argue that this operation is a reductive (indirect) generalisation of Σ to Σ'. Note that this is not a reduction of Σ' to Σ, but vice versa of Σ to Σ'. In the context of equivalences, however, this does not matter since, in addition to (4.48), the injectivity of q^{-1} with respect to Σ' holds, and the second line of (4.46) inversely becomes an embedding of Σ' in Σ. However, in addition to (4.48), q also has the property

$$\forall s.\Sigma(X;s) \rightarrow \exists t.\Sigma'(X;t) \wedge s = q^{-1}(X;t).. \tag{4.49}$$

of being *surjective*, and this makes the second line of (4.46) a candidate for a *refinement* of Σ to Σ', again with Σ' as the reducing theory. Of refinements, we will get to know their reduction qualities in the section after next. Again, in the case of an equivalence, the inversion holds that Σ is also a refinement of Σ', and so we are obviously faced with *degenerate* cases of embeddings and refinements. But this is precisely what justifies the preliminary treatment of these reductions unified in equivalences. In fact, one does not get ahead a single step in the reduction business without introducing equivalences as a special case of reductions from the outset.

Nonetheless, the conception of equivalences as reciprocal reductions gives rise to difficulties based on certain 'internal' asymmetries that can enter into (4.46). The requirements (4.46) and (4.47) explicitly state that the formation and crucial properties of the mappings q and q^{-1} may depend on Σ and Σ' respectively. Furthermore, it is at least not excluded that, conversely, the mapping q depends on Σ' or q^{-1} on Σ. Above all, however, in both cases it is admitted that one dependence, *but not the other one*, exists. This can, then, introduce the respective internal asymmetry into the equivalence. Equivalence (4.33)–(4.37) is a prime case in which such complications do *not* arise. However, let us consider the theory of the motion of a particle (a) in a potential field (Σ) and (b) in a conservative force field (Σ') as an example of the first case. Here, the transition from Σ to Σ' is achieved by a term q, which introduces the force field as the gradient of the potential. This q depends weakly on Σ insofar as its formation presupposes, for example, that the potential is differentiable. q, however, is completely independent of the decisive law of Σ: the equation of motion. In Σ', the latter is initially formulated for an arbitrary force field, and the decisive presupposition in Σ' is the *existence* of a potential field with a gradient that is the given force field. The formation of q^{-1} depends decisively on this existential presupposition, and we feel the asymmetry due to the ad hoc character of this presupposition, in contrast to the completely 'natural' formation of the gradient in the reverse case. The difficulty in dealing with such an internal asymmetry (q independent of Σ, but q^{-1} dependent on Σ', or vice versa) is, among other things, that it would suffice for mere symmetry if we had *in both cases dependence*—dependence of equal strength. Such a case is the reformulation of EG into terms of Riemann geometry mentioned at the beginning (cf. (4.56)–(4.60) of the following section). But it is impossible to make the *strength* of the dependence conceptually precise here, nor in general. In contrast, one could grasp the *mutual independence* in a more precise way by omitting the premises in (4.47), as well as in the claims (2.58) and (2.59) from Sect. 2.3, that q and q^{-1} concern internal terms. Moreover, the very possibility of *forming* q and q^{-1} should not depend on Σ or Σ', respectively. These conditions would make the independence absolute and transfer all 'gradual' differences into the dependencies. In any case, assessing symmetry in this case remains a matter that does not follow fixed rules.

To illustrate the other case—weak dependence of q on Σ' versus strong dependence of q^{-1} on Σ—, we choose the reduction

$$\rho = g(T) \wedge u = q(\rho) \vdash u = g'(T) \tag{4.50}$$

of the Stefan law g' (for Σ') to Planck's radiation law g (for Σ), which is considered in more detail in Sect. 4.4. Here, we seem to be dealing with a clear case. Planck's law, which concerns not only the temperature T but also the spectral energy density ρ, is a refinement of the Stefan-Boltzmann law, which concerns only the integral energy density u as a function of T. Accordingly, the transition from ρ to u through (the integral) q is highly degenerate, and q is, in particular, completely independent of g', i.e., Σ'. However, the degeneracy only holds as long as we disregard Planck's law. If we take it into account, the spectral densities are also different for different temperatures and thus, because of Stefan-Boltzmann, for different integral energy densities. With the definition

$$q^{-1}(u) \equiv g(g'^{-1}(u)) \,, \tag{4.51}$$

we then obtain the 'counter-reduction'

$$u = g'(T) \wedge \rho = q^{-1}(u) \vdash \rho = g(T) \tag{4.52}$$

to (4.50), which together with (4.50) means that the laws g and g' are equivalent. Of course, this succeeds only at a price that no physicist would pay, namely, the total dependence of q^{-1} on g (for Σ)—as opposed to the independence of q on g' (for Σ'). To accommodate our intuition here, which rejects (4.52) as a reduction, we are tempted to claim that, in principle, in an equivalence (4.46) and (4.47), the formation of q and q^{-1} be independent of Σ' and Σ respectively. As in the previous case, however, we have to keep in mind that a *dependence* would likewise suffice for symmetry and, therefore, we would again only arrive at intuitively assessing the cases. We should be aware of this fact when we include the internal symmetries in the concept of equivalence in the following.

From the problems discussed here, we have to distinguish another phenomenon, which can also only be grasped intuitively, but on the assessment of which not so much depends in a particular case. It concerns the difference in content of equivalent theories. In the two examples (4.33)–(4.37) and (4.38)–(4.45), the basic concepts transformed into each other are not very divergent. In other cases, the divergence can be considerable, and this will play a significant role in later chapters. For the moment, let us illustrate the matter by the example of two versions of *Euclidean geometry*. One is the version based on the congruency and betweenness relation we are already familiar with from (2.50)–(2.53) of Sect. 2.3. For the moment, it is not so important whether we choose the synthetic or analytical version distinguished at that time. This difference concerned the axioms, whereas the basic concepts were the same. This time, however, it is precisely the further reaching equivalence that is at issue, which includes *different* basic concepts. As an alternative basic concept, we now choose the concept of a distinguished coordinate system (within a certain set), more precisely here, of course: a global Euclidean coordinate system. In addition to the space M, a set $F \subseteq \mathrm{Pow}(M \times \mathbb{R}^3)$ of global coordinate systems on M is given in this axiomatic essentially belonging to the analytical method. With G as the Euclidean group of $\mathbb{R}^3$ (including dilatations), let hold for F that (1) with $\varphi \in F$ also $\theta\varphi \in F$ for all $\theta \in G$, and (2) $\varphi\chi^{-1} \in G$ for all $\varphi, \chi \in F$. So, this time, we have based our geometry entirely on the structure F (besides M), and this is certainly quite different from geometry with the congruence and betweenness relations as its foundation—*conceptually* different. Nevertheless, it now turns out that the two species of structure (for such are at issue here) are equivalent. Given F, one defines the relations cgr (congruence) and btw (betweenness) by (2.50) and (2.51) of Sect. 2.3, with $\varphi \in F$. This definition is independent of φ within F. Conversely, it follows from (2.50) and (2.51) as axioms that that the above axioms hold for F as a totality of coordinate systems in which cgr and btw have that Euclidean representation.

The difference in the basic concepts of otherwise equivalent theories, evident from the last example, gives rise to the following remark, which prepares the next section. From the definition (4.46) and (4.47) of our extended notion of equivalence, it is evident that it implies the ordinary equivalence

$$\Sigma(X; s) \wedge t = q(X; s) \dashv\vdash \Sigma'(X; t) \wedge s = q^{-1}(X; t) \,. \tag{4.53}$$

If, then, the equations in (4.53) were to be taken as definitions, our extended equivalence of Σ and Σ' could be regarded as the ordinary equivalence of certain definitional extensions of Σ and Σ' and, in this sense, could be traced back to ordinary equivalence. One is, then, tempted to ask what the point of extended equivalences in the sense of (4.46) and (4.47) might be. Apart from considerations on the status of the equations in (4.46) to be made below, we can say the following about this question.

Otherwise equivalent systems of basic concepts suggest *different possibilities of generalisation*, and that is why the differences between them should not be blurred by moving to (4.53). For instance, the path from coordinate-based Euclidean geometry to Riemann geometry is a much more direct one than that starting from the congruence and betweenness relations—indeed, one will hardly find any other path from the latter to Riemann geometry than via coordinate systems. Nonetheless, Riemann geometry is, of course, also a generalisation of Euclidean geometry in the version with the congruency and betweenness relations—however, only an indirect one, as addressed in the following section. Here, then, lies the reason for cultivating the differences in question: They may shed different light on one and the same thing in surprising ways and stimulate ever new theoretical possibilities (concerning not only generalisations).

At the end of this section, let us briefly address the status of the *equations* that appear in the formulae defining an equivalence (4.46). It suggests itself to regard these equations as *definitions* in the following sense: The question "what does 'T' mean in $t = q(X; s)$?" is to be answered by regressing to q as a complex term, as well as to the meaning of the 's'. The meaning of 't' is then extra-theoretical only insofar as the meaning of 's' is. As for the rest, it is determined intra-theoretically by q. In the sense of such definitions, we had understood the preparatory case (4.32), where in the Kepler-Newton example concepts such as 'orbital period' and 'velocity' are certainly defined concepts in the given sense, but both theories have the same basic concepts. The situation is different in case (4.46) and (4.47), where a complete exchange of the (typified) basic concepts can take place. Here, one has to take into account the inverse symmetry given by

$$t = q(X; s) \ \text{ and } \ s = q^{-1}(X; t) \,. \tag{4.54}$$

If the question "what does 't' mean?" is answered by referring to the left equation, then the question "what does 's' mean?" cannot also be answered by referring to the right equation since, otherwise, we would get stuck in a circle. One of the two equations, therefore, fails as an actual definition and must have some other status. The most we can say so far is: The sense of (4.46) may be that the s and t *can* be *equally* chosen as independent parameters and, then, the other *can* be defined in terms of them. But the choice of one of these two possibilities excludes the other. Even the attribution of the equivalence (4.46) to the ordinary equivalence (4.53) cannot deceive us about the fact that actually only *one* of the two equations may be used as a definition.

The associated 'symmetry breaking' corresponds to the fact that from (4.46) alone we cannot see which semantic function we want to give to the Eqs. (4.54). The question of what semantic role (apart from mutual definition) one can give them has been discussed in general for the "conditions of connectability" (Nagel 1961, Chaps. 11) or "bridge laws" (Hempel 1966, Chap. 5; Sklar 1967; Causey 1977, Chap. 2). We are in a similar situation with the equivalences (4.46) and (4.47) and still want to discuss the view of the Eqs. (4.54) as *synthetic identities*. The example of the following self-equivalence is particularly apt to make the situation clear. We make Euclidean geometry in the version of (2.50) and (2.51) (Sect. 2.3) slightly more explicit by adding to the dramatis personae, in addition to M, cgr and btw, a coordinate system φ for which just (2.50) and (2.51) holds. Let us assume to make this with another coordinate system χ in addition to φ. What is the relationship between the axiom systems $\Sigma(M; \text{cgr}, \text{btw}, \varphi)$ and $\Sigma(M, \text{cgr}, \text{btw}, \chi)$ obtained in this way? With τ as the Euclidean transformation that transforms φ into χ, the equivalence

$$\begin{aligned} &\Sigma(M; \text{cgr}, \text{btw}, \varphi) \wedge \chi = \tau \cdot \varphi \vdash \Sigma(M; \text{cgr}, \text{btw}, \chi) \\ &\Sigma(M; \text{cgr}, \text{btw}, \chi) \wedge \varphi = \tau^{-1} \cdot \chi \vdash \Sigma(M; \text{cgr}, \text{btw}, \varphi) \end{aligned} \tag{4.55}$$

of type (4.46) and (4.47) holds. This is a particularly clear case where it is just not obvious to interpret the associated equations as definitions. For *all* coordinate systems that are in play here are to be formed according to the following content-related rule: Establish a one-to-one mapping between points in space and triples of numbers in such a way that the relations cgr and btw, established as usual by transporting rigid rods, satisfy conditions (2.50) and (2.51). In particular, φ and χ in (4.55) are meant in this way, and there is nothing at all to define in this procedure but only to find the transformation τ that transforms φ into χ. The transformation equations (4.55) then express an *overinterpretation* of Euclidean geometry provided with several Euclidean coordinate systems. In fact, *one* such coordinate system is sufficient as the basis for a complete interpretation of the Euclidean terms. Every geometric fact formulated in a further (admissible) coordinate system also finds its expression in the first coordinate system, and if it concerns a theorem of Euclidean geometry, even the *same* expression. One *could* even introduce all the others from a single empirically established Euclidean coordinate system φ through Euclidean transformations. The question "what is χ?" for any χ would then be answerable by reference to the empirically obtained φ and an appropriate numerical Euclidean transformation τ. However, we immediately see that this is an extremely artificial view: Given the *uniform* external rule for establishing any Euclidean coordinate system, singling out one of them as the basis for defining all the others seems completely arbitrary. The appropriate view is that all coordinate systems are introduced externally and independently of each other by one and the same rule.

Of course, in general, we do not have such symmetrical relations as in the case of self-equivalence (4.55). We have already discussed this above for certain 'internal' asymmetries. If we have an arbitrary set of pairwise equivalent axiom systems for an empirical theory, there is no question of a uniform external interpretation of all respective typified basic terms $s, t, \ldots$. Even with only two equivalent axiomatisations as in the previous examples, nothing at all follows from the meaning of 's' and 't' about a possibly existing independent interpretation of 's' or 't'. In particular, it may be that the s are appropriate empirical terms for testing the theory, whereas the t are not. The only point that can always be made here is: If 's' or 't' is interpreted externally, then an independent interpretation of 't' or 's' is *not necessary*. One *could* take the path of definition, and the disadvantage of symmetry breaking would be countered by the advantage of not having to accept semantic overdetermination and related consistency problems. On the other hand, the actual occurrence of many theory equivalences in physics suggests that there is more than just definitions at play here. Of course, the occurrences of equivalences need not have interpretive reasons, and if they do, they need not have clear-cut or complete interpretations behind them. In a rough analysis, it can be said that, at the beginning of theory formation, physicists introduce the greatest possible variety of concepts without already knowing their relationships. Towards the end of theory formation, an axiomatic stage begins in which the tendency towards an economy of concepts and thus towards definitions predominates. This historical variation further complicates assessing the situation—not to mention the reluctance of theoretical physicists to engage in precise semantic regulation. To consider the Eqs. (4.54) *within* theory equivalences (4.46) as synthetic identities is accordingly not intended to claim more than that they concern empirically substantial statements—statements, however, with a content that does not result from establishing relations between semantically independent magnitudes in the usual way, but rather from over-interpreting the conceptual inventory of a theory in a way that is in principle avoidable but, nevertheless, not unwelcome. Even in everyday life, it is pleasant to be able to speak of both the evening star and the morning star, even though they refer to the same star.

4.3 Indirect Generalisations (Embeddings)

At the beginning of the previous section, we motivated the discussion of equivalences by their occurrence in indirect generalisations according to the scheme (4.29). The historically most important example is, as indicated, the generalisation of Euclidean geometry to Riemann geometry, and later, correspondingly, the generalisation of Minkowskian geometry to Lorentzian geometry (=the theory of Lorentz manifolds). We introduce the consideration of indirect generalisations with a reminder of the first-mentioned example. We first accept Euclidean geometry as a distance geometry

$$d : M^2 \to \mathbb{R}, \;\; d(x, y)^2 = \Sigma_i (x_i^\varphi - y_i^\varphi)^2 \tag{4.56}$$

as in (2.2)–(2.8) of Sect. 2.1. Here, φ belongs to the class of Cartesian coordinate systems. In Riemann geometry, on the other hand, we start with arbitrary differentiable and local coordinate systems. Besides F with

$$F \subseteq \text{Pow}(M \times \mathbb{R}^n) \tag{4.57}$$

as an atlas of such, we then have mainly a 2-fold covariant, positive definite and symmetric tensor field

$$ds^2 : TM^2/M \to \mathbb{R}, \;\; ds^2 = \Sigma g_{ik}\xi^i \eta^k \tag{4.58}$$

on TM^2, restricted to pairs with the same base point. Now, in what sense is this Riemann geometry a generalisation of Euclidean geometry?

As already said in our introductory remarks in Sect. 4.2, we are dealing with the direct generalisation of a theory equivalent to Euclidean geometry. We obtain the equivalence transformations in (4.46) here as follows. For a given Euclidean space $\langle M; d \rangle$, let F be the totality of (local) coordinate systems which are differentiable connected with a Cartesian coordinate system marked by (4.56), right side. Let ds^2 be that differential form which in a Cartesian coordinate system changes into the form (4.56), right side. The length of the straight line between the points x and y according to the Riemann differential form then coincides with the Euclidean distance d. In order to also get the inverse of q, we need to know a condition that restricts the tremendously general Riemann geometry to the Euclidean: the condition γ from (4.4) to (4.6). At this point, we can again see clearly how such conditions provide insights of different degrees into the reductions. What does it matter that a Riemannian space is Euclidean? A trivial (though successful) condition is that for all $x \in M$

$$g_{ik}(x) = \delta_{ik} \tag{4.59}$$

holds in at least one global coordinate system. On the other hand, it was an important and far-reaching insight that a vanishing curvature tensor

$$R^i_{klm} \equiv 0 \tag{4.60}$$

in a global coordinate system also leads back to Euclidean geometry. If one of these conditions is available, one obtains the Euclidean distance d as the distance according to (4.56), right side, in a global coordinate system with (4.59) or (4.60), and this is, then, the inverse of q. Since the curvature tensor is a defined term, strictly speaking we face a hybrid of (4.32), (4.46) and (4.47). In any case, the generalisation then takes place in the second step simply by omitting (4.59) and (4.60).

Since Euclidean geometry is a categorical theory, i.e., any two of its models are isomorphic (for a given spatial dimension), one usually thinks of one of its models as distinguished—just as *the* Euclidean space. Therefore, in the example of an indirect generalisation just given, the embedding character of such a generalisation does not become so clear. Nonetheless, it is of course true that in the given assignment we have assigned different Riemann spaces to different (albeit isomorphic) Euclidean spaces: The indirect generalisation embeds the class of more specific structures into that of more general ones. A clearer example in this respect, which also plays a role in physics, is given by measures as generalisations of functions—measures and distributions. The mathematical analysis of quantum mechanics showed, as is well known, that for position and momentum and, more generally, for all observables with a continuous spectrum, Hilbert space did not provide any eigenfunctions. Dirac, therefore, came to propose a formalism that (in the position representation), in addition to plane waves as eigenfunctions of the momentum, also provided so-called δ-functions as eigenfunctions of the position. However, these δ-functions were not functions in the usual sense. Mathematical analysis reconstructed them as distributions, and, as such, they were later introduced into axiomatic quantum field theory (cf. Bogolubov et al. 1990, Chap. 2, where the distributions are also called 'generalised functions'). In fact, already in classical statistical mechanics, in addition to real probability densities represented by functions in the usual sense (e.g. for the canonical distribution) δ-like measures appeared, such as the measure of the microcanonical ensemble which is δ-like for the (specified) energy surface. It is just that the aspect of generalisation was not so much emphasised here (from a historical point of view).

But here, too, we have something like that: Positive density functions are generalised to measures if the functions are sufficiently smooth. Let us consider this case in a little more detail (cf. Halmos 1950, Sect. 23). On a topological measure space $\langle X; \mu \rangle$, we consider a function f with properties such as

$$f \text{ continuous}\,, \quad f(x) \geq 0, \quad \int_X f(x)dx = 1\,. \tag{4.61}$$

To such a probability density, via

$$\nu(E) = \int_E f(x)d\mu \quad (E \subseteq X)\,, \tag{4.62}$$

we can now assign a measure ν, which is defined on a Boolean lattice of subsets E of X and, in turn, implicates certain properties, e.g.,

$$\begin{aligned} &\nu(E \cup F) = \nu(E) + \nu(F) \ \text{ for } \ E \cap F = \emptyset \\ &\nu(E) \geq 0, \quad \nu(X) = 1\,. \end{aligned} \tag{4.63}$$

This assignment is an *embedding* in the sense that (4.62) is injective, provided the functions F are sufficiently smooth, e.g., continuous:

$$\text{if } \int_E f(x)d\mu = \int_E f'(x)d\mu \text{ for all } E, \text{ then } f = f' . \tag{4.64}$$

With (4.61)–(4.64), we clearly have a genuine generalisation of the species of structure $\langle X; \mu, f\rangle$ by the species of structure $\langle X; \mu, \nu\rangle$ if we leave it at such general conclusions for ν as listed in (4.63). For, then, every function on X is characterised by a measure ν, but there are numerous measures to which no function (on X) corresponds, e.g., measures which only take the values 0 and 1 (probability functions on the subsets E). This generalisation, thus, extends our notion of generalisation of the first kind from Sect. 4.1, insofar as we have the embedding (4.62) *without* the measures corresponding to functions also *being* functions (on X). Nonetheless, here, the functions (4.61) are in a certain sense *reduced* to the measures (4.63). Of course, this is not a generalisation or reduction of the *concept* of function. For measures are also functions. What is generalised is the concept of a function *on* X.

However, the reduction obtained in this way is not elementary: It can be decomposed into the embedding of the set of functions (4.61) into the set of measures *to which functions from* (4.61) *correspond*, and a direct generalisation (in the sense of Sect. 4.1) of this set of measures to the set of *all* measures according to (4.63). Now this very decomposition shows that we are dealing with an *indirect* generalisation. It does not simply extend a concept, but first characterises it in a different way and then adds a direct generalisation. We summarise this again as in Sect. 4.1 (4.29) in the scheme

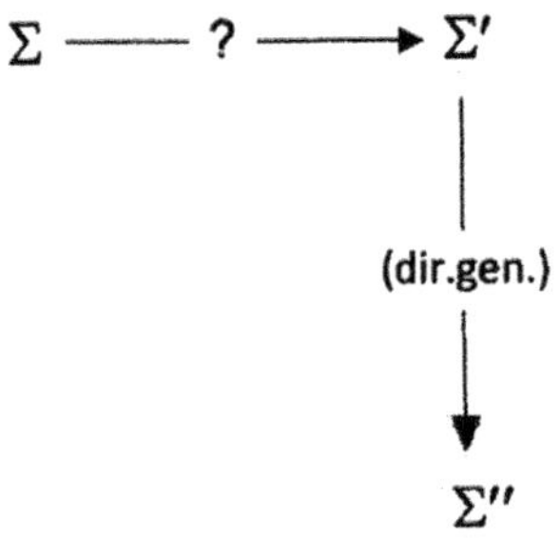

(4.65)

now indicating, however, that the question of whether the first reduction can be identified with an equivalence still needs to be clarified. We start to do so with a general consideration.

What is the first reduction in (4.65)? It seems obvious to generalise (4.62) by

$$\Sigma(X; s) \wedge t = q(X; s) \vdash \Sigma'(X; t) , \tag{4.66}$$

and (4.64) by

$$\Sigma(X; s) \wedge \Sigma(X; s') \wedge q(X; s) = q(X; s') \vdash s = s' . \tag{4.67}$$

Finally, we have to exclude the possibility that a direct generalisation, which according to (4.65) should only take place in the second step, is already involved in the first reduction. This is done by

$$\Sigma'(X;t) \vdash \exists s.\Sigma(X;s) \wedge t = q(X;s).. \tag{4.68}$$

On the other hand, it is easy to see that these conditions (4.66)–(4.68) already entail that Σ and Σ' are equivalent in the sense of (4.46 and (4.47). Crucial for this insight is the formation of

$$q^{-1}(X;t) \equiv \text{that } s \text{ with } \Sigma(X;s) \wedge t = q(X;s) \tag{4.69}$$

under the condition of Σ'. The existence in (4.69), then, follows from (4.68), the uniqueness from (4.67). If we have, more generally, at first only (4.66) and (4.67) before us, with Σ'' instead of Σ', we can *define* Σ' by the right side of (4.68) and thus get the decomposition into an equivalence of Σ and Σ', as well as a direct generalisation from Σ' to Σ''.

However, as was already discussed in Sect. 4.2, and as we will face again in the following section, we have to ask whether we want to accept the equivalence of Σ and Σ' obtained via (4.69) as such. Of course, this is an equivalence in the sense of (4.46) and (4.47). But we have, again, as in the previous section, the asymmetry that in (4.66) q can be independent of Σ', whereas g^{-1} in (4.69) always depends on Σ' because of (4.68). So we have the first of the two cases of internal asymmetry distinguished in the previous section. There, we already described the examples (4.33–4.37) and (4.38–4.45) as model cases in which internally everything is symmetric. We also encounter an ideal situation in topology with its various definitions as a species of structure (cf. the overview in Iyanaga and Kawada 1977, 408 B). For example, we can explain the topology on a space X sometimes in terms of open sets, sometimes in terms of closed sets. If we start from the open sets, the closed sets are explained as those with an open complement and, correspondingly, vice versa by interchanging ‘open’ and ‘closed’. Our case, transferred to this inversion, would mean that we instead introduce the open sets as those from which we obtain the closed ones as their complements. This is highly unsatisfactory, and, accordingly, we had already decided in the previous section not to admit such cases of equivalences (4.46) and (4.47) as equivalences *in a narrower sense*.

Of the two examples given so far in this section, the geometric one should satisfy the normal requirements for an equivalence. The situation is different with the second example. In the terminology of (4.65), for Σ' we have so far only the statements that ν is a measure and that a function f with (4.61) and (4.62) exists. Accordingly, for a definition of q^{-1} we have to rely on the form (4.69), and thus we have no equivalence proper. However, this becomes different if we replace Σ' by the equivalent statement (in the narrowest sense of equivalence) that ν is an absolutely continuous measure. In this way, the first line of (4.46) is trivially satisfied, and for the inversion we can use the theorem of Radon and Nikodym: Every (w.r.t. μ) absolutely continuous measure is representable in the form (4.62) (cf. Halmos 1950, Sect. 31). Then, we may insert into the second line an extensionally equal but intensionally different inversion of q, which defines f as the function that is constructed from ν in the proof of that theorem. This restores symmetry. (However, a condition to ensure the continuity of f would still have to be added.) To what extent this restoration is generally possible is hardly to be answered. It is, therefore, recommended to admit for the first factor of our decomposition, in addition to a proper equivalence, also asymmetrical equivalences in the disputed sense as *genuine* embeddings. In Sect. 5 (4.111)–(4.119), we will get to know further important cases of indirect generalisations which are problematic in this respect.

We have to provide for the concept of embedding as a concept of a reduction as indecomposable as possible for yet another reason. It already arises in such simple relationships as those of Newtonian particle mechanics to configuration space mechanics and again of the latter to general phase space mechanics. In both cases, the reason is that the reducing theory introduces other principal base sets than the one to be reduced. Therefore, the necessary embedding can no longer be done with an equivalence transformation as in the last example. (This also holds for the attempts to replace the usual (topological) space-time structures by algebras of functions, namely, to embed the former in the latter, cf. Earman 1989, Sect. 9.9.)

For Newtonian particle mechanics, we have above all the amount of particles and the potential energy U as a parameter, in addition to space and time. If we normalise the amount of particles as usual by the (ordinal and cardinal) number n of particles, we obtain the configuration space S^n (with S as space). Its differential structure is obtained by means of the Cartesian coordinate systems specified on S. From the masses m_k and the motions $\mathbf{x}_k$ of the particles, together with U, we obtain the Hamiltonian

$$H(\mathbf{x}_1, \ldots, \mathbf{x}_n; \mathbf{p}_1, \ldots, \mathbf{p}_n) \equiv \sum_k \frac{1}{2} m_k \cdot \mathbf{p}_k^2 + U(\mathbf{x}_1, \ldots, \mathbf{x}_n) \tag{4.70}$$

and the motion

$$\langle \mathbf{x}_1, \ldots, \mathbf{x}_n; m_1 d\mathbf{x}_1/dt, \ldots, m_n d\mathbf{x}_n/dt \rangle \tag{4.71}$$

in the phase space T^*S^n, i.e., in the cotangent bundle of S^n. From Newton's equations for U, one can finally derive Hamilton's equations for H. Correspondingly straightforwardly, one obtains general phase-space mechanics from the mechanics in configuration space. For a given configuration space M, we obtain the phase space as a cotangent bundle $T * M$. From the canonical linear form $\sum_i dp_i dq^i$, the symplectic metric ω on the phase space results by external differentiation in

$$\omega = \sum_i dp_i \wedge dq^i \ . \tag{4.72}$$

For a given Hamiltonian H, the Hamiltonian vector field leaving this form invariant is simply

$$\langle \partial H/\partial p_1, \ldots, \partial H/\partial p_n; -\partial H/\partial q_1, \ldots, -\partial H/\partial q_n \rangle \ . \tag{4.73}$$

The motion in configuration space is raised to phase space as in (4.71), and it follows from the Hamilton equations that we are dealing with the integral curves of the vector field (4.73) in phase space (cf. Westenholz 1978, Chap. 12).

It is intuitively clear that the two processes (4.70–4.71) and (4.72–4.73) concern embeddings—embeddings of the systems of the theory to be reduced into the systems of the reducing theory. Unlike in (4.66)–(4.69), however, we now admit a change of the principal base sets. The two decisive conditions are, thus, in general,

$$\Sigma(X; s) \wedge Y = P(X; s) \wedge t = q(X; s) \vdash \Sigma'(Y; t)\ , \tag{4.74}$$

with one-to-one correspondence

$$\begin{aligned} &\Sigma(X; s) \wedge \Sigma(X'; s') \wedge P(X; s) = P(X'; s') \\ &\wedge\ q(X; s) = q(X'; s') \vdash X = X' \wedge s = s'\ , \end{aligned} \tag{4.75}$$

where P and q are internal terms for Σ. Since, as the examples show, we also have very general frame theories of physics in mind here, we do not want to examine the behaviour of the cores in more detail. However, it is important to note that, as already in the special case (4.66)–(4.69), the reductions (4.74) and (4.75) are always decomposable into a *strict* embedding, which apart from (4.75) also satisfies

$$\Sigma'(Y; t) \vdash \exists Xs.\Sigma(X; s) \wedge Y \approx P(X; s) \wedge t \approx q(X; s).\ , \tag{4.76}$$

and a direct generalisation of Σ'. Since we now also admit a change of the principal base sets, we must no longer require the equality $=$ in (4.76) on the right, but only the isomorphism $\approx$. The latter automatically reduces to equality if the principal base sets remain fixed. A simple example of a strict embedding with a change of the principal base sets is obtained as follows: We consider structures $\langle X; s, \cap, \neg\rangle$, where s is a set of subsets of X, $\cap$ is the intersection and $\neg$ is the complement for subsets of X. If s is closed with respect to $\cap$ and $\neg$, then $\langle X; s, \cap, \neg\rangle$ forms a Boolean algebra, and it turns out that this is a strict embedding of the former species of structure in the latter. With regard to strictness (4.76), it is even true that every Boolean algebra is isomorphic to a perfect and reduced set algebra (cf. Sikorski 1969, Sects. 7 and 8).

Of course, the physical examples considered above do not deal with strict embeddings, but also with indirect generalisations in the sense of (4.74) and (4.75). Admittedly, this does not answer another question, with which we have dealt in more detail for example (4.61)–(4.64). There, introducing absolute continuity not only resulted in a 'symmetrical' equivalence with regard to the inversion of (4.62) but also gave a higher explanatory power to the subsequent direct generalisation (cf. Sect. 4.1). Restoring equivalence is no longer at issue in examples (4.70), (4.71), (4.72) and (4.73). But the character of the reduction would also be increased in these examples for both components if one could find corresponding conditions for a configuration space mechanics to be Newtonian or a phase-space mechanics to be a configuration space mechanics. However, it seems that nothing is known about such conditions.

Before we now move on from the generalisations, we would like to point out the explanatory condition γ of (4.4)–(4.6) (Sect. 4.1) once again. It is present in *all* generalisations concerning Σ—in the direct ones by definition and in the indirect ones—because these are always combinations of an equivalence or embedding and a direct generalisation. In Sect. 4.1, we have already emphasised this and how the γ has got lost in the orthodox concept of reduction. Since then, some authors have emphasised its indispensability (e.g. Yoshida 1977, 1; Hoyningen-Huene 1985), and even Nagel (1961, 434 and 442) has mentioned the matter again in passing. Its general significance is, on the one hand, that in crucial reductions the additional condition γ is often the 'missing link', the finding of which is the decisive step that might convince respective anti-reductionists. Today, it is often believed that, in biological systems, the laws of physics and chemistry are effective without any exception, without need of complementing them by biological laws proper in order to reduce biology to physics and chemistry. What is generally lacking, however, is an understanding of how to express in terms of physics the distinctive features that (generally speaking) make a physical system a biological one, *and* how, under these particular conditions, the specifically biological *concepts* emerge in which the theory to be reduced is usually expressed.

Our considerations of Sect. 4.4 immediately imply that, in a direct generalisation, every concept of the more general theory is also a concept of the more specific theory. Subsequently, it is easy to see that, conversely, the more specific theory offers possibilities of concept formation that are not available in the more general theory. For example, Kepler's theory Kep I–III from Sect. 4.2 works with concepts such as those of the semi-major axis and the orbital period, which in Newton's theory cannot be unconditionally defined because there, in general, the conditions under which the concepts in question are meaningful are not met: A hyperbolic orbit has no semi-major axis and no orbital period.

The general significance of this simple fact can hardly be overestimated. One of the main problems that arises in the 'great' reductions, e.g., of chemistry to physics or of biology to physics and chemistry, is the question: How should the concepts, the independent concepts, of these disciplines be at any rate obtained from the respective reducing discipline? Even for difficult reductions within physics, e.g., the reduction of thermodynamics to mechanics, this question arises seriously: How are concepts such as temperature and entropy to be obtained from purely mechanical concepts such as position, momentum, energy, etc.? This question can only be answered if we realise that such reductions essentially involve generalisations and that only the respective specific conditions of the reduced theory permit the introduction of these specific concepts.

4.4 Refinements

In a telling study of reductionism, Medawar (1974) used the classical hierarchy of geometries to explain the very hierarchy of sciences often cited with regard to questions of reductionism. The study is a typical example of how easily one can go astray in questions of reduction without having careful studies available. One may then invert the direction of reductions. This is what happens to Medawar, who compares the geometric hierarchy

1. topology
2. projective geometry
3. affine geometry
4. Euclidean geometry

with the hierarchy of empirical sciences that actually interests him:

1. physics
2. chemistry
3. biology
4. ecology/sociology

The comparison is based on thinking that, in both cases, we arrive at successively *more general* theories from the bottom up and, *therefore*, also have to look for reductions in this direction—for example, a reduction of biology to chemistry as a generalisation *like* a reduction of affine geometry to projective geometry, also in the sense of a generalisation. Now, it is clear that, in both cases, we ascend from the bottom up to theories that are successively more general in a vague sense. However, this does not at all imply that, for example, topology would be a generalisation of projective geometry and physics a generalisation of chemistry in roughly the sense in which, in this study, we have so far become acquainted with generalisations, and even understood them as reductions. Concerning the hierarchy of the sciences, we have to leave the question of their reducibility open here anyway. However, the geometries would only explain the hierarchy of the sciences in the present respect if, at least, they formed a hierarchy of genuine generalisations—or were, at least, reducible to each other from the bottom up in whatever sense. It seems, however, that the matter is rather the other way around and that generalisations are not at issue in either direction.

Let us illustrate this by means of affine and Euclidean geometry. The introduction of these geometries in the sense of Klein's Erlanger Programme is most quickly done with the help of a procedure already used in Sect. 4.2: A base set M (the space) is provided with a set F of global coordinate systems with respect to $\mathbb{R}^n$, which in one case is determined by the Euclidean, in the other by the affine group of $\mathbb{R}^n$; i.e., if G is this group, F consists of exactly those global coordinate systems $\varphi : M \to \mathbb{R}^n$ which emerge from one of them by applying a transformation from G. What is the relationship between these geometries? If $\langle M; F_{eu}\rangle$ is given, then F_{af} is uniquely determined by adding to F_{eu} all coordinate systems that can be obtained with transformations from G_{af}. Because of $G_{af} \supset G_{eu}$, for F_{af} determined in this way, $F_{af} \supset F_{eu}$ then also holds. More precisely, we have

$$\Sigma_{\mathrm{eu}}(M; F_{\mathrm{eu}}) \wedge F_{\mathrm{af}} = q(M; F_{\mathrm{eu}}) \vdash \Sigma_{\mathrm{af}}(M; F_{\mathrm{af}}) \tag{4.77}$$

if q is that mapping. It follows without further ado that, in addition,

$$\Sigma_{\text{af}}(M;\, F_{\text{af}}) \vdash \exists F_{\text{eu}}.\Sigma(M;\, F_{\text{eu}}) \wedge F_{\text{af}} = q(M;\, F_{\text{eu}})\,. \tag{4.78}$$

For with given F_{af}, choose any $\varphi \in F_{af}$ and let $F_{eu} = \varphi \cdot G_{eu} \cdot q$. Hence, q is surjective. Many Euclidean spaces have the same affine space in the sense of (4.77), and every affine space is the affine space of a certain Euclidean space in the sense of (4.78).

In this sense, Euclidean geometry is a *refinement* of affine geometry and the latter is a *coarsening* of the former. If we assume that the Euclidean or affine spaces of set theory were real physical possibilities, here we would obviously not be concerned with the case that the Euclidean possibilities are some of the affine ones or correspond to them by equivalence or embedding. Rather, it is crucial that for each of all the possible spaces, the Euclidean structure provides a finer, more informative description of space than the affine one. And it is, therefore, *not* that Euclidean geometry would be reduced to affine geometry, as would be the case if the former were more specific, but by (4.77) and (4.78) we reduce affine geometry as the coarser to Euclidean geometry as the finer. In the latter, the other is a part that can be 'deduced' within the framework of an ordinary theoretical development—not the other way around. Furthermore, Medawar is right when he claims "that as we go down the line [here: from affine to Euclidean geometry] the theorems and the concepts of the geometries become progressively richer and more particular", and it is also true that "as the geometries become more and more restricted, so new concepts 'emerge', concepts that simply do not appear in the geometry preceding it ..." (ibid., p. 61). But Medawar's point is supposed to be that this enrichment runs counter to the direction of reduction, as may be the case in the list of sciences. In the geometrical examples, however, it is the other way around, and we have to take note that richer conceptualisation sometimes can also appear in the direction of reduction.

Particularly misleading in Medawar's geometric hierarchy is that topology is at the top—indeed, perhaps it is precisely this that is to blame for turning the matter upside down, so to speak. As general as topology is, it is not a generalisation of Euclidean geometry, nor is the latter a refinement of the former. At best, topology is a generalisation of *Euclidean topology*, i.e., the topology induced by the Euclidean metric, and this, in turn, is a coarsening—i.e., a converse reduction of the Euclidean metric. There is, therefore, no clear direction of reduction between topology and Euclidean geometry. In passing, we want to mention a system of refinements in the framework of topology that is similar to the refinements in the framework of geometry and can serve as a further illustration (cf. Querenburg 1973, Chap. II). Richer (or: finer) than topological structures are the uniform structures, and again richer than these are the (topological) metrics. To a uniform structure canonically belongs a topology (always on the same space), and to a metric canonically belongs a uniform structure (and thus, a fortiori, a topological one). However, it is neither the case that every topological structure can be obtained in this way from a uniform one nor that every uniform structure can be obtained from a metric. But for a reduction in the present context of refinements, it would be important that the reducing theory accounts for every model of the theory to be reduced. Only then can one, in principle, 'forget' the reduced theory and all its models. In the present case, therefore, we have to find restrictions. Examples are given by the following theorems:

(I) The topological space of a uniform space is a T_{3a} space, and every T_{3a} space is uniformisable, i.e., there is a uniform structure with a topology that is the given one (Querenburg 1973, Theorem 11.26).
(II) The uniform space of a metric space is Hausdorffian and has a countable fundamental system, and any uniform space with these properties can be metrised (Querenburg 1973, Theorem 11.32).

In our language we will have to express this as follows: The theory of metric spaces is a refinement of the theory of uniform spaces with the properties given by (II), and the theory of uniform spaces is a refinement of the theory of topological spaces with the properties given by (I). However, it should already be pointed out here that the terminology of 'refinement' and 'coarsening' primarily expresses the relation between the *models* of the two theories.

The geometrical examples of this prelude will already have made it clear how refinements differ from the generalisations considered so far. In the latter, the basic idea is the *extension* of the theoretical scope of a theory with an essentially *unchanged* description of the single systems. New systems are added and, relative to them, one realises the specificity of the old, more specific systems. In contrast, refinements deal with the *same* class of possible systems as their coarsenings, but each system is described *more precisely*. The recovery of the coarser theory then takes place by means of a mapping, which transfers the finer description into the coarser description for each system. We have already had such mappings in the equivalences of Sect. 4.2, but there they had (necessarily) inverses, which now no longer needs to be the case. Equivalences are degenerate refinements: each of the two theories involved is a refinement of the other. However, we have seen from the example (4.50)–(4.52) that equivalences with internal asymmetries are to be considered as 'semi-genuine' refinements, so to speak, cf. (4.98)–(4.100) below.

To return to physics, we will first explain our new kind of reduction using the well-known example of *micro-* and *macro-states*. This example stems from the kinetic theory of gases, but the idea can already be explained with a simple urn model. Let our system initially consist of some (distinguishable) balls and some (distinguishable) urns into which the balls can be distributed. If we know which balls are in which urns, we know everything one wants to know about such a distribution. Let the various microstates of the system be given by this information. If we pass from the knowledge of which balls are in which urns to the mere indication—as macro-states—of *how many* balls are in each urn, this is obviously to dispense with information. If we know the distribution, we also know the corresponding macro-state by determining these numbers from the distribution of the balls. Conversely, however, from a macro-state—from the mere numbers—we can by no means reconstruct the micro-state to which it belongs: The micro-states describe the system more precisely.

If we now reconstruct this example in terms of an analysis of the species of structure, we get the following. *On the one hand*, we have two principal base sets K and U for the balls and the urns, respectively, both finite. We further have a microstate in the form of a mapping (as a typified base set)

$$f \subseteq K \times U, \quad f : K \to U\,, \tag{4.79}$$

where $f(x)$ is the urn in which the ball x lies. This defines a species of structure Σ_{mi}. *On the other hand*, we have a principal base set U (of urns) and (as a typified base set) a mapping

$$g \subseteq U \times \mathbb{N}, \quad g : U \to \mathbb{N} \tag{4.80}$$

($\mathbb{N}$ the set of non-negative integers). With g, it can be stated how many things are in an urn. What kind of things we are dealing with can be left open for this species of structure Σ_{ma} of the macrostate. We only determine this by unambiguously assigning a model $\langle U; g\rangle$ of Σ_{ma} to each model $\langle K, U; f\rangle$ of Σ_{mi}: We take the same U and

$$g(y) = \mathrm{card}\{x \in K \,|\, f(x) = y\}\,. \tag{4.81}$$

So again, following the pattern (4.77) and (4.78), we have expressed the coarsening by a surjective mapping and the corresponding refinement by its (non-unique) inversion.

It is not difficult to illustrate our new kind of reduction by further physical examples not simply concerning the coarsening states. *Kepler's theory* of the motion of a planet in its Newtonian version is a refinement of *Kepler's third law*. We want to follow this reduction in its converse form, in which Kepler's third law appears as a coarsening of the Newtonian version of Kepler's theory. For the latter, we have space and time as a fixed core and, in Euclidean coordinates, the equation of motion

$$d^2\mathbf{x}/dt^2 = -\mu\mathbf{x}|\mathbf{x}|^{-3} \tag{4.82}$$

with the constraint

$$\frac{1}{2}|d\mathbf{x}/dt|^2 - \mu|\mathbf{x}|^{-1} < 0\,, \tag{4.83}$$

which excludes unbound states. We also include Kepler's constant μ (which essentially contains the solar mass) in the core of the theory. (As statements about the motion $\mathbf{x}$, Kepler's three laws are equivalent to (4.82) and (4.83), cf. Sect. 4.2.) What are the parameters of Kepler's third law? This law connects the length a of the semi-major axis of the orbital ellipse with the orbital period T via

$$a^3/T^2 = \mu/4\pi^2\,. \tag{4.84}$$

We are, thus, dealing with a simple theory in the sense of Sect. 3.2, and the two variables a and T have as their value sets the sets of spatial and temporal distances. If D_S and D_T are these sets of values, the first part of the coarsening is, thus, performed by projections

$$P(\ldots D_S \ldots D_T \ldots) = D_S \text{ resp. } D_T\,. \tag{4.85}$$

In addition, however, we now have the mappings

$$\begin{aligned} a(\mathbf{x}) &= \text{length of semi-major axis of } \mathbf{x} \\ T(\mathbf{x}) &= \text{orbital period of } \mathbf{x} \end{aligned} \tag{4.86}$$

with $a(\mathbf{x}) \in D_S$ and $T(\mathbf{x}) \in D_T$, which are not projections but introduce new magnitudes. It is then decisive that the (coarsened) law (4.84) follows from the original law (4.82) and (4.83) together with (4.85) and (4.86) *and* that every solution $\langle a; T\rangle$ of (4.84) is the image of a solution x of (4.82) and (4.83) under mapping (4.86).

It is intuitively clear that the description of our system by $\mathbf{x}$ is considerably finer (or: more complete) than that by a and T. This relationship is expressed above all by the fact that innumerable motions $\mathbf{x}$ have the same a and, thus, T values. This can already be seen from considering the automorphisms of both theories: If we apply a spatial or temporal Euclidean transformation to $\mathbf{x}$, we arrive at an $\mathbf{x}'$ with the *same* $\langle a; T\rangle$ values. Such transformations thus leave the $\langle a; T\rangle$ values *themselves* invariant, whereas for the $\mathbf{x}$ they only leave their laws invariant. Now, two motions $\mathbf{x}$ and $\mathbf{x}'$, which emerge from one another through one of these transformations, are physically equivalent in a certain sense: They can be distinguished 'externally', but not 'internally'. In the account of physics, however, such equivalences are not eliminated, and in this respect our transition from $\mathbf{x}$ to $a(x)$ and $T(x)$ is already a genuine coarsening. But even motions that are physically different in every respect are mapped by (4.85) and (4.86) into the same $\langle a; T\rangle$ values. This is the case for all motions whose ellipses have the same major axis but different minor axes.

Similar to Kepler's case, and even simpler, is the plane mathematical pendulum with the equation of motion

$$d^2\varphi/dt^2 + \omega^2 \sin\varphi = 0 \quad (\omega^2 = g/l) \tag{4.87}$$

(l the length of the pendulum, g the gravitational acceleration). A coarsening of the full motion φ as a function of time is obtained here—according to (4.85)—by retaining the value sets of angles and durations, and—according to (4.86)—by the terms

$$\begin{aligned} \alpha &= \text{Amplitude of } \varphi \quad (0 \leq \alpha \leq \pi) \\ \tau &= \text{Oscillation period of } \varphi \quad (\tau > 0)\ , \end{aligned} \tag{4.88}$$

for which from (4.87) results the relationship

$$\tau = 4\omega^{-1} \int_0^{\pi/2} (1 - \sin^2 \alpha/2 \cdot \sin^2 \beta)^{-1/2} d\beta\,. \tag{4.89}$$

As with the formation of (4.86), the possibility of forming the terms (4.88) is essentially based on the underlying theory (4.82) and (4.83) or (4.87). Only certain properties of $\mathbf{x}$ or φ guaranteed by the respective theories ensure the unambiguous existence of a and T, or α and τ. Incidentally, it is again true that many systems φ (of different phases) have the same amplitude (or oscillation period), i.e., there is a genuine coarsening and correspondingly a genuine refinement, since for given α and τ with (4.89) always an associated solution φ of (4.87) exists. Law (4.89) can thus be completely understood by law (4.87).

If we now ask for the notion of *refinement in general*, then the more mathematical examples first considered suggest connecting the reducing theory T' and the reduced theory T by a deduction

$$\Sigma'(X; s) \wedge Y = P(X; s) \wedge t = q(X; s) \vdash \Sigma(Y; t) \tag{4.90}$$

of the axioms Σ of T from the axioms Σ' of T', where P and q are internal terms with respect to Σ'. We had yet worked with such deductions in (2.57)–(2.62) of Sect. 2.3, but then the terms X and s were not intended to have physical meaning. In contrast, now we suppose that the $\langle X; s\rangle$ as models of Σ' are possible physical systems, for which P and q specify certain other structural moments Y and t for which $\langle Y; t\rangle$ are models of Σ according to (4.90) and which, as such, form new systems that are physically possible according to T. In addition to (4.90), for the previous examples, we had noted the 'inversion'

$$\Sigma(Y; t) \vdash \exists Xs.\Sigma'(X; s) \wedge Y = P(X; s) \wedge t = q(X; s). . \tag{4.91}$$

In (2.57)–(2.62) of Sect. 2.3, it already appeared in a similar form. Now, it guarantees not only that all T' systems explain a corresponding T system (according to (4.90)), but that also *all* systems considered possible by T are explained by an appropriate system of T'.

The requirements (4.90) and (4.91) for the concept of refinement would be fine, and perhaps even exhaustive, if all the structures claimed in (4.90) and (4.91) indeed stood for possible theory systems. But this is hardly ever the case: Although we never exceed the limit of what is possible to think, it is still always a long way to what can be considered *physically* possible. Accordingly, we had requested that the systems possible according to a theory all have a core firmly connected to the theory as a common fragment for which, moreover, an axiomatic of its own was expected. As in the previous cases of other kinds of reductions, we now have to ask how the core and its axioms enter into reductions by refinement. And, first of all, we have just to ask whether the previous claims (4.90) and (4.91) can be maintained in view of these limitations. There seems to be no problem with regard to (4.90). Of course, (4.90) is stronger than the requirement that is actually only meant and restricted by the core condition. Its formulation requires splitting q into a 'constant' part q (in new denotation) and a 'variable' part q_1. For we will require accordingly that the core $\langle X°; s°\rangle$ of Σ' passes into the core $\langle Y°; t°\rangle$ of Σ through the deduction mappings:

$$\vdash Y° = P(X°; s°), \;\; \vdash t° = q(X°; s°)\,. \tag{4.92}$$

To this we add for the core axioms:

$$\Sigma'^{\circ}(X^{\circ}; s^{\circ}) \vdash \Sigma^{\circ}(Y^{\circ}; t^{\circ}) . \tag{4.93}$$

(4.90) is then replaced by

$$\Sigma'^{\circ}(X^{\circ}; s^{\circ}, s) \wedge t = q_1(X^{\circ}; s^{\circ}, s) \vdash \Sigma(Y^{\circ}; t^{\circ}, t) , \tag{4.94}$$

where only s and correspondingly t are variables indicating the multiplicity of still admitted systems. With (4.92), then, (4.94) follows from (4.90) (with corresponding renaming) and would also be all we need of it physically. But it turns out that the stronger (and simpler) requirement (4.90) is satisfied in all applications. In this sense, it poses no problems.

The situation is different with the requirement (4.91). Due to the existence quantification occurring in it, it is too *weak* if we restrict the domain of possible systems by introducing cores. For then, of course, the existence claim should already hold with respect to these (fixed) cores. Let us consider the example (4.79)–(4.81) of micro- and macro-states. There, (4.90) is satisfied as long as all parameters are variables, i.e., no cores appear: For a given $\langle U; g\rangle$, we still have the choice of K and, thus, its cardinal number. And this is also necessary because for (4.91) we must satisfy the condition

$$\sum_{y \in U} g(y) = \mathrm{card}(K) ,$$

which follows from (4.81). If we now stipulate U and K to be constants U° and K°, respectively, and make $\langle U^{\circ}, K^{\circ}\rangle$ and U° the core of Σ_{mi} and Σ_{ma}, respectively, we can no longer satisfy this condition and, therefore, the physically relevant part of (4.91) either: f would have to be found on K° and not on any K, and that need not work. (The example is also relevant for the relations in μ space and Γ space of statistical mechanics to be considered later: the number of particles is basically a variable.) We now propose

$$\Sigma(Y^{\circ}; t^{\circ}, t) \wedge \Sigma'^{\circ}(X^{\circ}; s^{\circ}) \vdash \exists s.\Sigma'(X^{\circ}; s^{\circ}, s) \wedge t = q(X^{\circ}; s^{\circ}, s). \tag{4.95}$$

as a strengthening of (4.91) adapted to the existence of cores. This statement contains the required existential strengthening from mere quantification to a particular instance (here: X° and s°). On the other hand, it is also somewhat weakened by the newly added premise Σ'°. This premise anticipates the part of Σ' in the conclusion for which Σ cannot reasonably account. We will illustrate this situation in later examples.

Equations (4.93) and (4.94) thus take the place of (4.90) and (4.91) as the main basis for the concept of refinement. This does not mean that the implications (4.90) and (4.91) are forgotten, for they too are mostly met in the examples, (4.90) being too strong, (4.91) too weak. Under the aspects given by (4.90) and (4.91), it is also worthwhile to look back at the embeddings. In fact, we already used deductions to define the (strict) embedding in (4.74)–(4.76) of the last section. Then, however, the actual deduction took place in the opposite direction: Σ' was deduced from Σ and not Σ from Σ'. This made sense in conjunction with the requirement (4.75) of the *one-to-one correspondence* of the deduction mappings $\langle P; q\rangle$, which, in turn, aimed at their use as embeddings in (indirect) generalisations. Here, in refinements, the typical cases are precisely those in which the deduction mappings are *strongly degenerate*, i.e., many systems of the finer theory map into the same system of the coarser theory. This time, the essential additional condition is that the deduction mappings are *surjective*: (4.91) and even (4.95). Only in this way can any system of the coarser theory be captured by the finer one. And, only if this is guaranteed, can we speak of a reduction of the coarser theory. In contrast, the surjectivity required in (4.76) (up to isomorphism and again in the opposite direction) only served to split off strict embeddings as *indecomposable* reductions. The comparison of refinements with embeddings shows particularly well how little deduction *per se* matters in a reduction. What is crucial is how it is used.

This becomes clear once again by considering the deductions that are effective in both kinds of reduction and being aware that they take place in *different* directions: in the case of embeddings, in the direction of the reduction; in the case of refinements, in the opposite direction. As in the case of equivalences, we also have to realise here that the possible semantic function of a deduction is not unambiguously linked to the direction of reduction. In the case of equivalences, this was the case if the deduction functioned as a definition. For a definition cannot be inverted in one and the same context. We only recovered symmetry here in case of synthetic identities, as, for example, in coordinate transformations. In all the examples of embeddings and refinements given so far, but particularly in the latter, the semantic function of the deductions in question is now that of a *definition*. We learn the meaning and significance of the terms on the left side of the deductions through the terms on the right together with the meaning of their arguments. In this way, we do not face problems of symmetry in one and the same context here. For the genuine embeddings and refinements are uni-directional. But the above remark should say that, in principle, terms of the reducing theory as well as those of the reducing theory can both be defined by terms of the respective other reduction partner.

Although in the previous examples the deductions can always be understood as definitions from a semantic point of view without restraint, this semantic function is not *necessarily* connected with the embeddings or refinements (see below). Here, we again face Nagel's problem of inhomogeneous reduction, as already touched upon in Sects. 4.2 and 1.2. Concerning the definitions, however, the independence of a theory based on defined concepts is at stake and needs to be justified. For this circumstance seems to bind it inevitably to the theory from which the defining concepts are taken. Let us take, for example, the terms of the amplitude α and the oscillation period τ in the theory of the pendulum. The original theory (4.87) established a relation between the (full) motion φ and the limit frequency ω; the coarsened theory (4.89) does so for α, τ, and ω. Now the independence of the latter theory obviously cannot consist in the fact that we finally completely forget about the pendulum as a system and replace it by a new system that is only about α, τ, and ω. For which system could that be? Which one would do us the favour of behaving according to the same law (4.89) that we know from the pendulum and yet is not the pendulum? And, even if there were such a system, what would be the general significance of such a transition?

Compared to these imaginings, it is appropriate to consider the theory (4.89) as also being about the pendulum, but only giving a coarsened description of the pendulum compared to the original one. Such a coarsening can, in principle, even be known prior to the corresponding refinement, and in the case of the pendulum it was indeed the case. Amplitude and oscillation period (or better: frequency) have always been more readily accessible to measurement than the full temporal function φ. Even today, using my pendulum clock, I can, at least, empirically confirm the simplified relationship $\tau = 2\pi/\omega$ for small amplitudes without knowing anything about Eq. (4.87). Accordingly, although in a slightly different way, e.g., classical mechanics as configuration space theory can, in principle, be conceived on its own even before identifying it as a generalisation of Newtonian mechanics by defining, e.g., the configuration space or the Hamiltonian from Newtonian concepts. The independence of the former from these definitions becomes obvious here yet because there will, in general, be many models of the generalisation for which there are no corresponding models at all of the more specific theory that is to be embedded.

We now have to consider some pitfalls that the concept of refinement generally defined by (4.92)–(4.95) still has. We have already said at the beginning that our attempt to characterise the different kinds of reduction in general in no way claims completeness in the sense that *every* theory relation admitted by the attempted characterisation of a kind of reduction is indeed a reduction of this kind. We only want to, and can, give certain characteristics that suffice to distinguish reductions of different kinds, e.g., generalisations from refinements, and these again from asymptotic reductions, and so on. This restriction also applies to the refinements, but in contrast, e.g., to the generalisations, we can give concrete examples showing that with the claims (4.92)–(4.95) we certainly did not yet reach our goal. This criticism, however, only concerns the interpretation of the respective operations as refinements and not any other properties they may have as reductions.

Let us take as a *first* example the theory of the pendulum, however, restricting the solutions of (4.87) to *finite* time intervals. Let this theory be Σ'. Then, we correlate via q each such solution with the solution that it indeed uniquely determines for *all* times (together with Σ'). Let the latter (usual) theory be Σ. Together with Σ', it satisfies (4.92)–(4.95), but it cannot be said that Σ' refines Σ. On the contrary, the system descriptions of Σ' per se are significantly more incomplete than those of Σ. What is at issue here? In the counterexample, the solutions of Σ' are chosen in such a way that several of them can still be combined to describe a system, or: several can still be partial descriptions of the very same system. The fact that in (4.94) the same t can belong to several s, therefore, only expresses a refinement if the different solutions *exclude each other*, such as the different solutions of (4.87) for all times. This is a requirement of the completeness of a theory per se. It could, therefore, be a deficiency of our conception of a theory (and not only of the concept of refinement) if we admit incomplete theories in the sense indicated. But how should we exclude them? If our request is that different models of a theory also correspond to different physical systems, we have to realise that within *one* theory its systems are only represented by *its* models. But how should we know, from the models per se, that they exclude each other in the description of the system?

A *second* problem is connected with the question already raised in Sect. 2.3: whether we should not restrict the species of structure admitted a priori as physical axioms to those with means of expression entirely restricted to the respective core, in particular, with regard to the bound variables and their typifications. In the present context, this question arises as follows. Let us assume that we have a species of structure Σ' as the axiom of a physical theory, as well as a core $\langle X^\circ; s^\circ \rangle$, and so on. Then, we can try to find coarsenings of this theory by simply taking (as a vehicle) with respect to Σ' internal terms P, q, and q_1, for which P and q depend only on X and s, and q as well as q_1 are typified by P (relative to Σ'). Then, by

$$Y = P(X; s), \quad t = q(X; s), \quad t_1 = q_1(X; s, s_1) ,$$

we would already have a structure $\langle Y; t, t_1 \rangle$ for each $\langle X; s, s_1 \rangle$. And we would have a new core through

$$Y^\circ = P(X^\circ; s^\circ), \quad t^\circ = q(X^\circ; s^\circ) .$$

What is asked for is a species of structure Σ which completes these data in the sense of (4.90) and (4.91) or (4.92)–(4.95).

Let us first make sure that specifications as just indicated offer themselves quite easily in physical practice. Take the example of two interacting particles in classical mechanics. The (anyway) variable parameters of an associated theory of interaction are then the masses m_1 and m_2, as well as the motions $\mathbf{x}_1$ and $\mathbf{x}_2$ of the two particles. A clear case of coarsening would be to simply ignore one particle, i.e., to pass from the characterisation by $\langle m_1, \mathbf{x}_1; m_2, \mathbf{x}_2 \rangle$ to, say, $\langle m_2, \mathbf{x}_2 \rangle$. Another case would be the transition from a many-particle system

$$x = \langle q_1, p_1, \ldots, q_N, p_N \rangle \tag{4.96}$$

in phase space (Γ space) to the measure

$$\varrho_x(\Delta) = \text{Number of particles from } x \text{ in } \Delta \tag{4.97}$$

in 1-particle space (μ space), where Δ is an arbitrary (Borel) subset of this space. This too, like its simplification (4.79)–(4.81), is a clear case of voluntarily neglecting information, expressed by coarsening[1]: According to (4.96) and (4.97), many microstates x pass over into the same macrostate ϱ_x, just as in the previous example many characterisations $\langle m_1, \mathbf{x}_1; m_2, \mathbf{x}_2 \rangle$ of the 2-particle system pass over into the same characterisation $\langle m_2, \mathbf{x}_2 \rangle$ of the second particle.

The problem is, however, that these coarsenings usually do not result in a physically reasonable theory. In the first example, we had such a theory for the uninteresting case that the particles have no interaction. Then each particle had its own Hamiltonian, and we could simply take the Hamilton equations with its Hamiltonian as the

[1] "This only holds for fixed cell division, cf. vol. II, p. 140." (Handwritten comment in E. Scheibe's copy of the book, referring to Scheibe 1999.).

law for the second particle. But, in the case of a genuine interaction, none of the particles would have its own dynamics (by definition). And the same is true without exception for the measures ϱ_x, given the dynamics of the microstates. Of course, in both cases, we could describe the motion for the coarser structure by taking the original, finer characterisation. We could say $\langle m_2, \mathbf{x}_2 \rangle$ is of the kind that a $\langle m_1, \mathbf{x}_1 \rangle$ *exists* such that both together satisfy the given interaction. And, correspondingly, we could say that a measure ϱ changes because an x *exists* such that $\varrho = \varrho_x$ and $\varrho(t) = \varrho_{x(t)}$, where $x(t)$ changes in phase space according to the Hamiltonian dynamics. Indeed, this procedure is generally at our disposal: As long as the axioms Σ are not subject to any restrictions of the kind indicated above, we can probably always find an Σ in the form of the right-hand side of (4.91) for the given data such that not only (4.90) and (4.91) but also (4.92)–(4.95) is satisfied. But, obviously, in both examples, we do not arrive at any reasonable dynamics in this way, and the general procedure also seems questionable because of the existence constraint. For it incorporates into the coarsening elements of the finer theory which affect the independence of the coarsening. The axioms of the refinement must not enter directly into the coarsening. In particular, it must be possible to know the coarsened (and, thus: reduced) theory *prior* to each of its refinements.

The conditions described above thus face us with the choice of being more or less liberal in admitting species of structure as physical axioms and thus, correspondingly, obtaining, more or less, reductions by refinement. In principle, here, it would be preferable to pursue the problem of a further, drastic restriction of physical species of structure. The corresponding limitation of possible refinements need not frighten us. For there is no fundamental interest in ensuring that every coarsening of a physical theory (in the sense of given mappings P, q, etc.) in turn results in a meaningful physical theory. In practice, on the other hand, it can be advantageous to take the more liberal path. In particular, coarsenings of the kind indicated in our examples occur in the context of approximate reductions and can be used to advantage for clear analysis if they are themselves understood as reductions. Indeed, under certain conditions, approximations to the structures coarsened in this way have their own dynamics, and their reduction to the original dynamics can be clearly decomposed if the coarsenings are available as an exact component, cf. Chap. 4 of Vol. II.

Finally, we turn to a *third* peculiarity that can occur in refinements. It reveals an aspect under which our concept of refinement fixed by (4.92)–(4.95) (or already (4.90) and (4.91)) does not seem too broad but, instead, too narrow—too narrow, at least, if formal equivalences in the sense of Sect. 4.2 are, at most, understood as a degenerate case of refinements. Formally speaking, in an equivalence each theory is a refinement of the other. However, we had already seen in Sect. 4.2 that certain asymmetries can be hidden in our equivalences of theories, e.g., that the term q in (4.46) can be formed without knowing and using Σ', but not q^{-1} without knowing and using Σ. Even more: For $\langle X; s \rangle$ which are not subject to Σ, q can be a highly degenerate mapping, but restricted to models of Σ, the inverse q^{-1} clearly exists. We had already pointed this out by (4.50)–(4.52) and suggested in the case of such an asymmetry not to speak of equivalences, *even though* (4.46) and (4.47) applies.

Thus we can (cf. Schpolski 1954, Sect. VI) reduce the Stefan–Boltzmann law

$$u = \alpha \cdot T^4 \ \ (\alpha = 1.08 \cdot 48\pi k^4/c^3h^3) \ , \tag{4.98}$$

with u and T as the energy density and temperature of cavity radiation to Planck's law

$$\varrho(\nu) = 8\pi/c^3 \cdot \nu^3\{\exp(h\nu/kT) - 1\}^{-1} \tag{4.99}$$

in the sense of deducing (4.92)–(4.95) from (4.98) via

$$u = \int_0^\infty \varrho(\nu)d\nu \ . \tag{4.100}$$

The same applies to Wien's displacement law

$$\lambda_{\max} \cdot T = \beta \ \ (\beta = 1/5 \cdot hc/k) \tag{4.101}$$

(most simply) with the λ version of Planck's law

$$\varrho'(\lambda) = 8\pi c \cdot \lambda^{-5}\{\exp(hc/kT \cdot \lambda^{-1}) - 1\}^{-1}, \tag{4.102}$$

if one defines λ by

$$\lambda = \text{Wavelength for which } \varrho' \text{ has its maximum.} \tag{4.103}$$

Any physicist will recognise these processes as reductions or explanations of (4.98) and (4.101) by Planck's law. In particular, *each* of the transitions from ϱ to u in (4.100) and from ϱ' to λ_{max} in (4.103) *per se* is a high information loss, and the deductions are surjective in the sense of (4.95). So far, nothing prevents Planck's law from being understood as a refinement of both the Stefan–Boltzmann law and Wien's displacement law. On the other hand, however, both mappings (4.100) and (4.103) can be inverted *if* restricted to the solutions of Planck's law. Then different ϱ also belong to different T, and different T in turn belong to different u or λ_{max}, respectively. From this point of view, we are no longer talking about a genuine refinement. Nevertheless, no physicist would claim that the inversions in question had now explained Planck's law, be it from Stefan-Boltzmann's, be it from Wien's. The reason for denying this might be that, in contrast to the very general formulae (4.100) and (4.103), their inversions can only be obtained under very specific conditions (i.e., Planck's law). In particular, it is no wonder that, conversely, one obtains these conditions back by explanation. In general, one will also have to pay attention to considering the deduction (4.94) *independently* of Σ as a refinement, based on the behaviour of the mappings $\langle P; q, q_1 \rangle$.

With the above example, we finally have a case of refinements in which the deductions are not definitions from a semantic point of view. Admittedly, they would also be in the present case if the two equations of (4.98) and (4.101) in brackets were understood as definitions of α and β, respectively. But, then, the two main Eqs. (4.98) and (4.101) would not really be Stefan-Boltzmann's and Wien's law, respectively. For these, the two constants α and β have been *empirically* determined at a time when their connection with the universal constants c, h, and k (and even h itself) was not yet known. Only with Planck's law came the possibility of *explaining* α and β. For this explanation, however, the equations in question are just not to be understood as definitions of α and β. Rather, the right-hand side connects the empirically known left-hand side with universal constants, reducing it to them. The equations are, therefore, empirical insofar as the connection of the measured α and β with the constants c, h, and k could have turned out to differ from the given ones. In this case, no reduction of (4.98) and (4.101) to Planck's law would have come about. On the other hand, the two equations in question only concern semantic differences in determining the same matter; they do not concern a purely external relationship between physical magnitudes as in the case of usual empirical laws.

Concerning the still missing 'balance of success' for the refinements, it is crucial—to anticipate this right away—that the deduction equations be *valid*, no matter whether due to their character as definitions or to an empirical status. In case of the latter, as in the last example given for the constants α and β, then, as we have just seen, already derivability in the sense of (4.94) depends essentially on the validity of the derivation equations; if they are empirically valid, their justification depends on the same corroboration mechanism as assumed for the theory partners in a reduction (cf. Sect. 3.1). The validity in question thus presupposed, we next face the question of what about the inheritance of confirmations and successes from Σ to Σ', if we assume (4.92)–(4.95). Concerning the operation mode of (4.95), it is guaranteed that confirmations and successes of Σ are not destined to fail because they were achieved for a system that realises a solution of Σ, *but not of* Σ'. In the direct generalisations we had the stronger condition (4.4), but (4.95) also suffices.

As far as conservation itself is concerned, we face a similar problem as in the case of direct generalisations. We had to extend the concept of corroboration of Sect. 3.1, more precisely: its domain of application, in order to achieve that all confirmations and successes of Σ, are also those of Σ'. The same is needed this time too, although for a different reason. Then Σ' proved to be logically too weak to account for all successes of Σ, now it is terminologically too weak: Confirmations and successes of Σ are expressed in a language—the $\langle Y^\circ; t^\circ, t\rangle$ language of (4.92)–(4.95)—which is not the $\langle X^\circ; s^\circ, s\rangle$ language of Σ'. To express observation statements of Σ and Σ' only *elementary* statements of Σ and Σ' were admitted, so that already the correctly expressed observation statements of Σ are no such statements for Σ'. Nevertheless, successes of Σ must also be successes of Σ. For Kepler's theory (4.82) and (4.83), the empirical validity of Kepler's third law (4.84) is, of course, a success, and likewise for the pendulum theory (4.87), the empirical validity of (4.89). Hence, we have to achieve that empirical successes of (4.84) or (4.89) are also empirical successes of (4.82) and (4.83) or (4.87), and we can do so very easily by simply *requiring* so. At this point, this looks rather ad hoc—as did the corresponding requirement for the direct generalisations in Sect. 4.1. However, it is not, if, from the point of view of the reducing theory Σ', we consider the following.

The restriction of the observation statements and data to the elementary statements and terms of a theory Σ' is quite unjustified, given that, in general, also many *defined* concepts of a physical theory are usually considered to be good empirical terms. For a pendulum, for instance, α and τ in (4.88) cannot be measured according to (4.87) merely following their definition, so to speak, and *calculating* them from an already empirically determined motion $\varphi(t)$ and ω. The simpler procedure is rather to *measure* α and τ *directly* and to determine just thereby, at least partially, φ, which is much more difficult to access. And exactly the same applies to α and T with respect to the planetary motion $\mathbf{x}(t)$, according to (4.86). Here, as well as there, we have to be aware: Although in both cases we do not determine a magnitude by following its definition, we can be sure to measure via the direct method also what we wanted to measure according to the definition. So, whenever we encounter such relations, it is obvious that the corroboration cases of a theory Σ' should be extended in the following sense: If some concepts $q_1(X^\circ; s^\circ, s)$ of Σ' (in the sense of Sect. 2.4) are gathered to form a new theory Σ such that (4.92–4.95) holds, and if these concepts are empirical in the sense just indicated, then the confirmations and successes of Σ formulated in these terms are also confirmations and successes of Σ'.

By having thus indicated how to extend the corroboration and refutation cases for a theory Σ', the question of whether we can exactly demarcate those terms of Σ' for which the extension is possible still arises. Such a demarcation would then also affect the concept of refinement, in so far as further conditions on the terms q_1 (and possibly q) in (4.92)–(4.95) came about. Although there is much to be said about this, it does not seem that this specification is possible in a satisfactory way (cf. Ludwig 1990, 106–110). What can be done, of course, is to investigate how the testing possibilities improve or worsen by passing from concepts to other concepts via internal terms q_1, according to Sect. 2.4. Let us take, for example, the transition (4.36) from the spectral energy density ρ as a function of the frequency to the same as a function of the wavelength of the radiation. This case was presented earlier as a prime example of an equivalence, namely, of Planck's law (4.33), for ρ as a function of frequency and the same law (4.34) for ρ' as a function of wavelength. The faultlessness of this case, which actually consists only of two different formulations of one and the same law, is based not least on the fact that perfect symmetry prevails here also in the empirical respect: Every observation statement for ρ with respect to (4.33) corresponds unambiguously to an observation statement for ρ' with respect to (4.34) and vice versa, i.e.:

$$\rho(\nu^\circ) = \rho^\circ \text{ means } \rho'(\lambda^\circ) = \rho'^\circ \quad (\text{with } \rho^\circ = c\nu^{\circ-1} \text{ and } \rho'^\circ = \rho^\circ c^{-1}\nu^{\circ 2})$$
$$\rho'(\lambda^\circ) = \rho'^\circ \text{ means } \rho(\nu^\circ) = \rho^\circ \quad (\text{with } \nu^\circ = c\lambda^{\circ-1} \text{ and } \rho^\circ = \rho'^\circ c\lambda^{\circ-1})\,.$$

This, insofar as neither of the two theories is distinguished against the other—provided that it is possible to determine frequencies and wavelengths equally well.

Another case is the transition (4.100) from the spectral density ρ to the integral u in the same theory. Here, we have an improvement of the empirical conditions insofar as the observation statements about ρ are of second order, but those about u are of first order: u itself is an empirical datum, whereas for ρ only the pairs of values $\langle \rho^\circ, \nu^\circ \rangle$ with $\rho(\nu^\circ) = \rho^\circ$ are empirical. The price for this improvement, however, is such an observation statement about ρ has no equivalent for u, and u can even not be considered determined for any finite set of observation statements. A similar conclusion apparently holds for (4.85), (4.86) and (4.88). From these and many other examples, we see that from an empirical point of view, the logical and set-theoretical tools used to define a concept can only help to a certain extent to determine the concept according to its definition. Sometimes they are just an obstacle as, e.g., in the geometrical definition of a straight line on the basis of the betweenness relation. Whether one of three points lies between the other two may be empirically decided in certain cases. But whether this holds *for any three points* of an arbitrary set of points can usually no longer be decided. From this point of view, the concept of a straight line is, then, hardly an empirical concept. It may not be possible here to separate the still tolerable cases from the unacceptable ones without arbitrariness.

Next to the preservation of confirmations and successes guaranteed as above, we have to note that, for the refinements, it is not possible to transform failures into successes—in contrast to the direct generalisations. Extending the domain of application $\mathfrak{R}$ of the coarser theory T to include such successes of the refined theory T' which were failures of T is of no use, given that always $\mathfrak{R}' = \mathfrak{R}$, strengthening C_1 from Sect. 3.3. So, to improve the situation, no new systems are sought, but rather more precise descriptions of the old ones. Accordingly, every failure of T is also a failure of T'. From a logical point of view, this is immediately obvious on the basis of (4.94), if, as requested at the beginning, we absolutely insist on the validity of the deduction equations. New successes, therefore, always concern only the refinement *as such*, i.e., as far as it goes beyond the conditions of the coarser theory. Then, the description of these successes does not permit any retranslation into the language of the coarser theory, and just this is its novelty.

4.5 Extensions and Improper Generalisations

Mathematical logicians call a (first-order) theory T' an "extension" of a (first-order) theory T if the non-logical symbols of T are also non-logical symbols (of the same syntactic category) of T' and, correspondingly, all theorems of T are also theorems of T'—in short, if the language of T is a part of the language of T' and T formally derives from T' (cf. Shoenfield 1967, 41). For any model of T', then, its T-fragment is a model of T, i.e., T also semantically derives from T', and conversely T formally derives from T' if the latter is the case. However, it is something special if all theorems of T expressed only in the language of T' are already theorems of T, and correspondingly if every model of T is a fragment of a model of T'. We then speak of syntactically and semantically 'conservative' extensions, respectively. Every semantically conservative extension is also syntactically conservative but, in general, not vice versa. In the following, we will transfer the concept of a semantically conservative extension to species of structure.

This would first result in the formal implication

$$\Sigma'(X, X'; s, s') \vdash \Sigma(X; s)\,, \tag{4.104}$$

where the s are typified by the X in Σ' in the same way as in Σ, and the partial language condition was directly adopted in the notation. Further, the condition of (semantic) conservativity would have its counterpart in the "inversion"

$$\Sigma(X; s) \vdash \exists X's'.\Sigma'(X, X'; s, s'). \tag{4.105}$$

and can, thus, be formulated in purely syntactic terms here, where set theory is already built into the theories. Again, Σ is then not just any, but the strongest formal consequence of Σ', expressed only in the X and s, i.e., every such consequence already derives from Σ ('syntactic' conservativity). Conservative extensions in the sense of (4.104) and (4.105) are now nothing but certain special cases of refinements in the preliminary sense of (4.90) and (4.91). They arise if we choose *projections* for the deduction terms P and q. In these, the $P(X; s)$ are simply some of the X, and the $q(X; s)$ are some of the s. To put it more precisely:

$$P_i(X; s) \equiv X_{\lambda(i)}; \quad q_k(X; s) \equiv s_{\mu(k)}\,. \tag{4.106}$$

The X and s *not* chosen in this way appear as X' and s' in (4.104) and (4.105).

Now, however, the conditions (4.90) and (4.91) were not yet decisive as a general description of refinements. To take the cores and core axioms into account, we had to proceed to (4.92)–(4.95). Our notion of extension, if we want to obtain it as a special case of (4.92)–(4.95), must therefore be adapted to these conditions. This is done by the approach

$$\langle X^\circ, X'^\circ; s^\circ, s'^\circ\rangle \text{ resp. } \langle X^\circ; s\rangle \tag{4.107}$$

for the cores of T' or T, and

$$\Sigma'^\circ(X, X'; s, s') \equiv \Sigma^\circ(X; s) \wedge s' \in \sigma'^\circ(X, X') \wedge \alpha'^\circ(X, X'; s, s') \tag{4.108}$$

for the core axioms. From the other two conditions (4.94) and (4.95), we obtain

$$\Sigma'(X, X'; s, s', s_1, s_1') \vdash \Sigma(X; s, s_1)\,, \tag{4.109}$$

as well as

$$\Sigma(X^\circ; s^\circ, s_1) \wedge \Sigma'^\circ(X^\circ, X'^\circ; s^\circ, s'^\circ) \vdash \exists s_1'.\Sigma'(X^\circ, X'^\circ; s^\circ, s'^\circ, s_1, s_1').\,. \tag{4.110}$$

Extensions in the sense of (4.107)–(4.110) may at first seem particularly attractive to those who want to see logical implications at the centre of reductions, given that in (4.104) and (4.110) no additional premises intrude, in contrast to (4.90) and (4.94). However, we should bear in mind that, here, as in (4.92)–(4.95), we only make a reduction claim if at least (4.110) is also satisfied, and that, therefore, in the case of equal languages (omission of the primed terms), the extensions pass over into equivalences, and the "pure" implication as a non-trivial relation is thus, once again, eliminated. We draw attention above all to two cases.

There is, first of all, a widely known case in which extensions play a non-trivial, if somewhat hidden, role—a double role, to be precise. The question is whether, and in what sense, for example, the theory of the linear harmonic oscillator defined by the equation of motion

$$m \cdot d^2q/dt^2 = -k \cdot q \tag{4.111}$$

is a special case of the general mechanics governed by the equation

$$m \cdot d^2q/dt^2 = K(q)\,. \tag{4.112}$$

The same question can be asked, for example, in the more substantial case of Lorentz's electron theory, given that, there, the (non-relativistic) equation

$$m \cdot d^2\mathbf{x}/dt^2 = -e(\mathbf{E} + 1/c[d\mathbf{x}/dt, \mathbf{H}]) \tag{4.113}$$

is obviously a special case of the more general equation

$$m \cdot d^2\mathbf{x}/dt^2 = K(\mathbf{x}, d\mathbf{x}/dt)\,, \tag{4.114}$$

which also admits velocity-dependent forces. And the Schrödinger equation

$$i\hbar \cdot \partial\psi/\partial t = -\hbar^2/2m \cdot \nabla^2\psi - e^2/r \cdot \psi \tag{4.115}$$

for an electron in the Coulomb field—a third example—appears as a special case of the more general equation

$$i\hbar \cdot \partial\psi/\partial t = -\hbar^2/2m \cdot \nabla^2\psi + V(r) \cdot \psi \tag{4.116}$$

with arbitrary potential V. Here, we have the most significant case in which a central parameter, left open in a general theoretical framework, is specified in one way or another.

That one is so easily inclined to think of the relation between the special and the general in all these cases is simply due to the fact that, for the more general right-hand side of the Eqs. (4.112, 4.114, and 4.116), one can *choose* the corresponding right-hand side of the Eqs. (4.111, 4.113, and 4.115). And yet, for a more detailed analysis, we have to take into account two issues. *First*, the Eqs. (4.111, 4.113, and 4.115) themselves are not the specifying equations, but only derive from the latter together with (4.112, 4.114, and 4.116). Rather, the equations that specify (4.112, 4.114, and 4.116), the γ conditions of Sect. 4.1 (4.5) and (4.6), are, to make it quite clear,

$$K(q) = -k \cdot q \tag{4.117}$$

and

$$K(\mathbf{x}, \mathbf{v}) = -e(\mathbf{E}(\mathbf{x}) + 1/c[\mathbf{v}, \mathbf{H}(\mathbf{x})]) \tag{4.118}$$

and

$$V(r) = -e^2/r\ , \tag{4.119}$$

respectively. The deductions of the Eqs. (4.111, 4.113, and 4.115) from (4.112, 4.114, and 4.116) and (4.117)–(4.119) are no longer a specifying step but, rather, the inversion of extensions, and this is the point where extensions enter into the reduction at issue of (4.111, 4.113, and 4.115) to (4.112, 4.114, or 4.116), respectively—and, indeed, *themselves* enter as a reduction step. *Second*, there is the difficulty that each of the Eqs. (4.111, 4.113, and 4.115) include parameters with meanings that do not belong to the general conceptual armoury of classical mechanics or quantum mechanics. For example, the concepts of charge, as well as electric and magnetic field intensity, are concepts of classical electrodynamics, and one would not like to consider electrodynamics or even Lorentz's theory a special case of classical mechanics, *despite* the suggested subordination. Although classical mechanics provides a general scheme for the behaviour of forces as causes of motions, it does not teach us what kind of forces there are in the world, and, thus, there is no path leading from it to **E** and **H**. To quantum mechanics, these classical fields are even more alien. In other words, it is only possible after an extension that is significant as regards content, albeit formally insignificant, to specify directly the respective general theory represented by the Eqs. (4.112, 4.114, and 4.116). This extension, which introduces the electric and magnetic fields, e.g., is *not* itself a reduction step in the whole procedure.

Let us now elucidate the details for the case (4.113), (4.114) and (4.118). The claim is simply that the Lorentz theory (4.113) reduces to mechanics (4.114) by an extension and a subsequent 'improper generalisation'.[2]

The Lorentz theory, centred on Eq. (4.113), consists mainly of space and time (restricted to an inertial system for the non-relativistic case), motion $\mathbf{x}$, mass m, and charge e of a particle (all variable), as well as the (also variable) fields **E** and **H** with associated field intensity spaces. In addition, the Eq. (4.113) enters as the physical main thing. This theory is extended in a first step by a force space and a (variable) force field K (possibly velocity-dependent), as well as mainly the Eqs. (4.114), (4.118), *omitting* (4.113). The resulting theory is immediately shown to be an extension of the Lorentz theory in the sense of (4.107)–(4.110), in particular, of (4.110). In a second step, we then simply omit Eq. (4.118). This is a direct generalisation, though not to plain mechanics, but to what may be called a (formally insignificant) *improper extension* of it. For the theory also includes charge and the electrodynamic fields, however, without any significant condition but simply as ontologically independent entities that cannot be obtained from mechanics. Thus, this second step is a direct generalisation, not to mechanics proper but to an insignificant generalisation of it:

---

[2] Scheibe uses the term *'Fast-Verallgemeinung'* to denote a kind of reduction that is *almost* a generalisation, or *comes close to* a generalisation. We translate it here by 'improper generalisation' in order to indicate its connection to an insignificant extension (*unwesentliche Erweiterung*), for which we correspondingly use the term 'improper extension' (see below).

an *improper generalisation*, as we shall say, to mechanics proper. Altogether, the procedure follows the scheme

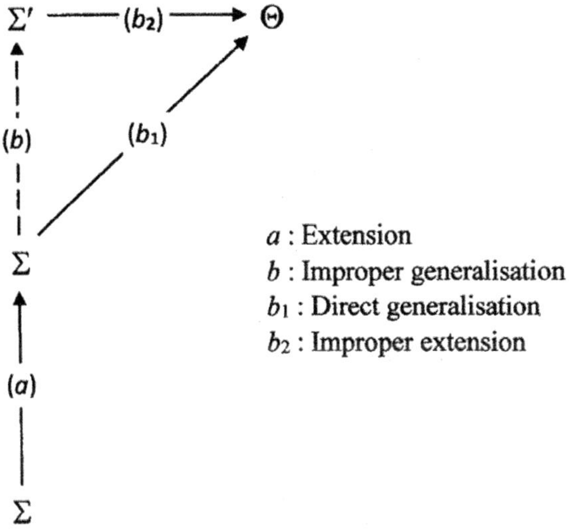

(4.120)

which illustrates how Σ is reduced via Σ' to Σ''. Without introducing a *new* kind of reduction, we, of course, also have a reduction of Σ (via Σ') to Θ. However, we are not concerned here with the relationship between Σ and Θ but between Σ' and Σ''.[3]

The concept of an improper extension, which is decisive for that of an improper generalisation, will not be further specified here. It should be noted that it would be the *inverse* of an extension, which, in an improper generalisation (if we decompose it), would succeed the direct generalisation (cf. the arrow directions of (b_1) and (b_2) in (4.120)). However, the inversion of an extension is certainly not a reduction, and this is precisely why the improper generalisation is a new kind of reduction—if it is a reduction at all. That an extension is improper is meant here only in the following sense: The entities newly added to the extension, in the above example the charge and fields added to general mechanics, are at best connected to each other but not coupled to the theory that they extend. It is like introducing a topology on a group but saying nothing about how it is related to the group operation (e.g., by the continuity of the latter on the former). Similarly, in our example, mechanics is extended by electrodynamic entities but without any coupling as given by (4.118). It was precisely this coupling that had taken place by (4.118) in the extension, which was our first reduction step. In the extension (b_2) of (4.120) there is nothing like this. Hence, for an improper generalisation, we can speak of a reduction to mechanics insofar as this extension is improper in the above sense. In other, ontological respects we can not, as already indicated. It is not realistic to assume that the actual forces of nature are already given, let alone known, by the general concept of force as available in

[3] In Scheibe's example, Σ is the Lorentz theory, Σ'' is mechanics proper, and Θ is an improper extension of the latter, namely, mechanics plus charge and electromagnetic field.

mechanics. Insofar as Maxwell's equations of electrodynamics specify how fields arise from material sources, electrodynamics is not reducible to mechanics anyway. Quite different are the Eqs. (4.113), (4.114) and (4.118), which state how the fields in the special case of a point-like charge carrier react on the latter. Here, the term 'Lorentz force' already says that (4.118) is about a special force, and this is what brings mechanics into play. However, the restriction expressed by calling the resulting reduction 'improper' is a significant ontological reservation.

The subsumption of the theory relations (4.111)–(4.119) under the scheme (4.120) and, thus, in particular, the use of extensions in (4.120 a) could face the objection that these cases concern no genuine refinement according to (4.106), since according to the Eqs. (4.117)–(4.119) different parameters of the respective right-hand side also correspond to different forces or potentials on the left-hand side. Although we have always admitted such boundary cases, in the present case it is worthwhile to have a look at another possible analysis. Indeed, we encounter here the remarkable case that a theory T is *reducible in two different ways* to another theory T'. This was not ruled out by any of the previous settings, and we now face such a case. What is more, this case is not an arbitrary one. Rather, it is a guiding idea for the presentation of extensive parts of physics, such as classical mechanics, quantum mechanics, continuum mechanics, etc. Their development always takes place based on specifications such as those present in the Eqs. (4.117)–(4.119). The fact that (4.117)–(4.119) are all about *injective*, one-to-one mappings, i.e., precisely what makes these cases non-genuine refinements in (4.120), now draws our attention to the other route to embeddings or indirect generalisations (4.66)–(4.69), discussed in Sect. 4.3, for which this one-to-one correspondence is constitutive.

Indeed, in (4.111)–(4.119), the Eqs. (4.111, 4.113, and 4.115) are quite specific laws of motion, which according to (4.66) and (4.67) of Sect. 4.3 are embedded by (4.117)–(4.119) into rather general theories centred around (4.112, 4.114, and 4.116). Unlike the previous interpretation, in which these equations belong to the intermediate theory Σ' of scheme (4.120), they are now to be read as the transformations $t = q(X; s)$ of (4.66)–(4.67) which make the *transition* from the reduced theory (4.111, 4.113, or 4.115) to the reducing theory (4.112, 4.114, or 4.116). In this regard, we should again raise the question already discussed in Sect. 4.3, whether the embeddings by (4.117)–(4.119) cannot be decomposed into an equivalence on the basis of (4.68) and a subsequent direct generalisation. The present examples, however, are precisely ones in which the reversal (4.69) based on (4.68) alone leads to a rather 'crooked' equivalence: The formation of, e.g., the Lorentz force in (4.118) does not depend in any way on the assumptions of the initial theory Σ in (4.66)–(4.69), whereas a reversal according to (4.69) enforced by (4.68) depends significantly on Σ'. Thus, decomposability is only given here in the sense of decomposing into a genuine embedding and a subsequent direct generalisation, which is always possible.

Extensions, as they occur in the context of the generalisation problem just discussed, are not completely trivial in the following regard. The relation between the species of structures Σ and its extension Σ' is not simply

$$\Sigma'(X, X'; s, s', s_1, s_1') \equiv \Sigma(X; s, s_1) \wedge \dots , \tag{4.121}$$

with Σ as a conjunction member in Σ'. For the core species of structure, we have demanded just this particular relation because it does not seem to be a significant restriction with regard to the applications. For the complete axiom systems, it would be a restriction, as the examples (4.111)–(4.119) show. But (4.121) occurs frequently nonetheless—indeed, the construction of more complicated theories actually always proceeds in successive steps from extension to extension according to (4.121): What is to be said about $\langle X; s, s_1 \rangle$ is complete with Σ, and now it goes on to the extension by X', s', and s_1. Thus, for example, a sequence starts from Euclidean geometry of space; continues with the geometry of space and time (by the way, an improper extension in Newton's approach); succeeded, based on space and time, by a dynamics that specifies forces; and, finally, by means of the latter (and space and time), an equation for the motion of matter is established.

(By the way, the last extension is conservative due to the requirement that the equations of motion have solutions for every admitted dynamic.) Thus mechanics is a (conservative) extension of Euclidean geometry, and so is quantum mechanics. The point of view of refinement is even well brought out here, insofar as the object to be altogether described in a mechanical theory is still very incompletely described by the mere description of space. Nonetheless, we would not like to consider Euclidean geometry to be *reduced* to classical mechanics or quantum mechanics. But we would also, perhaps, not like to say that Euclidean geometry *is deduced* from Newtonian mechanics or quantum mechanics. And, yet, it *is* in a, albeit trivial, sense. Being disappointed about this only means not yet having understood something correctly: For the construction of these theories, as we have in mind here, it is not at all to be expected to make any use of the central mechanical assumptions of the theories for that reduction or deduction. With another theory construction, which does not include the concept of space from the outset, this just might turn out to be the case, and then one might rightly be surprised (cf. Weizsäcker et al. 1958). Here, however, we are simply faced with the ever-present and unavoidable phenomenon of degenerate cases appearing *somewhere* at the boundaries of our concepts.

The *definitional extensions* also belong here. They even form a subclass of (4.121) and have the yet more specific form

$$\Sigma'(X; s, s', s_1, s_1') \equiv \Sigma(X; s, s_1) \wedge s' = q(X; s) \wedge s_1' = q_1(X; s, s_1) \,, \tag{4.122}$$

where for the core species of structures holds

$$\Sigma'^{\circ}(X; s, s') \equiv \Sigma^{\circ}(X; s) \wedge s' = q(X; s) \,, \tag{4.123}$$

with q and q_1 as internal terms with respect to Σ° and Σ, respectively. Thus, Σ' arises from Σ by explicitly defining certain terms of Σ (cf. Sect. 2.4), namely, the $q(X; s)$ and the $q_1(X; s, s_1)$. Here, the $q(X; s)$ are already terms of Σ°, and the core species of structure Σ'° arises by definition from the core species of structure Σ°. Another interpretation of formulae (4.122) and (4.123) is that the s'° (together with X° and s°) and the s_1' (as variables) are externally interpreted. Then the Eqs. (4.122) and (4.123) are not definitions but synthetic identities, as already met at the end of

Sect. 4.2. One would then have 'reinterpreted' terms of Σ° and Σ. Either way, we are dealing with trivial extensions, since Σ' and Σ, as already Σ'° and Σ°, in fact, are equivalent in the sense of Sect. 4.2, Eqs. (4.46) and (4.47).

4.6 Unifications (Multiple Reductions)

Already in Sect. 1.1, voices from physics were echoed that express the desire for a unity of physics—unity, of course, at the same time with a certain completeness. A very simple mechanism for advancing this goal is *multiple reduction* in the sense that several theories $T_1, T_2, \ldots$ are reduced to one and the same, further theory T. According to our minimal conceptions of the notion of reduction developed in Sect. 1.3, such multiple reduction is a process of unification of theories, insofar as, in a fully reduced complete inventory of physics, only T needs to be listed of the theories in question, whereas $T_1, T_2, \ldots$ can be 'deduced' from T on the basis of well-established analytical methods—namely, by the procedure that is converse to the respective reduction. The multiple reduction does not immediately lead to a deeper insight into what constitutes the 'inner' unity of a theory, here of the theory T, although it does at least demonstrate what 'reduction potential' T has, given that $T_1, T_2, \ldots$ are reducible to T. On the other hand, we do not have to fear the well-known triviality argument that a multiple reduction can already be had by choosing T as a conjunction of $T_1, T_2, \ldots$ (cf. Hempel 1965, 273, Note 33, where precisely this is put forward as a difficulty of explaining laws as well). For mere logical deducibility is not eo ipso a reduction in our view as has been repeatedly emphasised. It is rather to be hoped that the reduced theories show their significant interdependence just through a multiple reduction. Admittedly, we cannot yet unfold the cases of possible multiple reductions at this point since the approximate reductions are not yet available to us. But the two historically perhaps most famous unifications of theories can, indeed, only be achieved by exact reductions—however, the second one with a blemish.

The first of these cases is the unification of electrostatics, magnetostatics, and optics in Maxwell's fundamental equations of electrodynamics. We will first formulate these three theories separately, and then show how they can be reduced to the fourth. It should be noted from the outset that electrostatics and magnetostatics are not Lorentz-invariant, so the reduction requires the introduction of specified coordinate systems that are transformed into each other only by translations and spatial rotations. Electrostatics deals with a time-independent electric field **E** and a likewise time-independent charge distribution ρ, for which in the specified coordinate systems the field equations

$$\mathrm{rot}\mathbf{E} = 0, \quad \mathrm{div}\mathbf{E} = 4\pi\rho \tag{4.124}$$

and the constraint

$$\partial\mathbf{E}/\partial t = 0 \tag{4.125}$$

hold. In magnetostatics, a magnetic field **H** and a current density **j** add, both also time-independent, with the field equations

$$\begin{aligned} \mathrm{rot}\mathbf{E} &= 0, & \mathrm{div}\mathbf{H} &= 0 \\ \mathrm{rot}\mathbf{H} &= 4\pi/c \cdot \mathbf{j}, & \mathrm{div}\mathbf{E} &= 4\pi\rho \end{aligned} \tag{4.126}$$

and the constraints

$$\partial\mathbf{E}/\partial t = \partial\mathbf{H}/\partial t = 0, \tag{4.127}$$

all in specified coordinate systems, too. Furthermore, in optics we deal only with the fields **E** and **H**, with the associated Lorentz-invariant field equations

$$\begin{aligned} \mathrm{rot}\mathbf{E} &= -1/c \cdot \partial\mathbf{H}/\partial t, & \mathrm{div}\mathbf{H} &= 0 \\ \mathrm{rot}\mathbf{H} &= 1/c \cdot \partial\mathbf{E}/\partial t, & \mathrm{div}\mathbf{E} &= 0\,. \end{aligned} \tag{4.128}$$

In contrast to (4.124)–(4.128), full electrodynamics finally deals with a system of **E** and **H**, as well as ρ and **j**, with the Lorentz-invariant Maxwell equations:

$$\begin{aligned} \mathrm{rot}\mathbf{E} &= -1/c \cdot \partial\mathbf{H}/\partial t, & \mathrm{div}\mathbf{H} &= 0 \\ \mathrm{rot}\mathbf{H} &= 1/c \cdot \partial\mathbf{E}/\partial t + 4\pi/c \cdot \mathbf{j}, & \mathrm{div}\mathbf{E} &= 4\pi\rho\,. \end{aligned} \tag{4.129}$$

Concerning the intention to reduce the three theories (4.124)–(4.128) to (4.129), we first note that they are equipped with different basic concepts, but only in such a way that, in (4.124) and (4.125), **H** and **j** are missing and, in (4.128), ρ and **j**, whereas in (4.126) and (4.127) all the basic concepts of (4.129) are already present. So, for the intended multiple reduction, the first step will be to *extend* (4.124), (4.125) and (4.128). Equation (4.124) and (4.125) is extended by **H**, **j**, and the equations

$$\mathbf{H} = \mathbf{j} = 0\,, \tag{4.130}$$

whereas (4.128) is extended by ρ, **j**, and the equations

$$\rho = \mathbf{j} = 0\,. \tag{4.131}$$

In both cases, we mean extensions in the technical sense of this term (cf. previous section), although only improper extensions in the sense that the newly added terms are not connected to the already existing ones. Even more, the vanishing of the respective magnitudes expressed in (4.130) and (4.131), of course, justifies omitting them at all (in the opposite direction).

In a second step, it now follows without further ado that (4.129) is a *direct generalisation* of (4.124)–(4.128), (4.130) and (4.131)—now *including* the latter equations. Indeed, the following equivalences hold:

$$\begin{aligned}(4.129) \wedge (4.130) &\equiv (4.124\text{-}4.125)\\(4.129) \wedge (4.127) &\equiv (4.126\text{-}4.127)\\(4.129) \wedge (4.131) &\equiv (4.128)\ .\end{aligned} \tag{4.132}$$

Here, (4.124)–(4.126) and (4.130) are restricted to specific coordinate systems. Moreover, magnetostatics is a direct generalisation of electrostatics:

$$(4.126\text{-}4.127) \wedge (4.130) = (4.124\text{-}4.125)\ . \tag{4.133}$$

We thus have transitivity, in that magnetostatics is a generalisation of electrostatics and the Maxwell equations of magnetostatics. This would result in (4.130) ∧ (4.127) as additional premise for the composite reduction. But, in this case, (4.130) alone is sufficient. Whereas (4.125) and (4.127) each introduce a rest frame, (4.131) and thus also (4.128) are Lorentz-invariant. This is immediately apparent from the Lorentz-invariant notation

$$\begin{aligned}\eta_{\lambda\mu|\nu} + \eta_{\mu\nu|\lambda} + \eta_{\nu\lambda|\mu} &\equiv 0\\ \eta^{\lambda\mu}{}_{|\mu} &\equiv 4\pi/c \cdot j^{\lambda}\end{aligned} \tag{4.134}$$

of Maxwell's equations (cf. Rindler 1960, Sect. 42). Full electrodynamics is thus a direct generalisation of optics, and this reduction is Lorentz-invariant. It is a combination of an extension and a direct generalisation of the other two theories (4.124), (4.125), (4.126) and (4.127), and these reductions (like the theories) are not Lorentz-invariant.

The other classical case of unifying independently developed theories is the unification of Kepler's laws with Galileo's law of falling bodies by Newton's theory of gravitation. We would not deal with this case now if Galileo's law were understood as his quantitative law of falling bodies and Newton's theory as its strict multi-body version. For, then, the issue would only work with approximate reductions. If, however, Galileo's law is understood to mean only the statement that 'all bodies fall with equal velocity', and Newton's theory is understood to mean the approximate Eq. (4.30) from Sect. 4.3, we arrive at the following consideration: Kepler's theory (Kep I–III) from Sect. 4.2 is a direct specialisation of (4.30) with the additional premise (4.31). Whereas, here, (4.30) refers to the sun as the centre of gravity, the historically correct Galilean statement is obtained by referring (4.30) to the earth. For, from (4.30), it follows that two masses dropped at the same distance from the centre of the earth fall in the same way (except for a rotation) and, in particular, have the same acceleration. This qualitative result is based solely on the equivalence principle and the central symmetry of the gravitational force. Hence, it has to be noted that *this* statement itself is much too general to look generalised or otherwise reduced by the approximate version (4.30) of Newton's theory. We only obtain a proper reduction in our sense by taking the quantitative form of Galileo's law. However, the resulting reduction is then only approximate (cf. Sect. 5.3). There is yet another issue that we cannot adequately address here, nor later: Kepler's laws, on the one hand, and Galileo's law of falling bodies, on the other, are only representative of the traditionally separate domains of physics of the celestial spheres and sublunary physics, and their unification in one and the same theory is also only representative of the grandiose step of overcoming this fatal separation of two areas of reality.

The topic of the unification of theories in the previous sense also includes the remarkable subsumption of many particular and quite specific physical theories under such general formalisms as classical mechanics, quantum mechanics, or a general (classical or quantum) field theory. Here, i.e., in (4.111)–(4.114) of the previous section and, in general, in all 'special cases' found in a textbook of mechanics, we are literally dealing with multiple reductions if we commit ourselves to one of the two alternative kinds of reduction distinguished in Sect. 4.5; and the mentioned general framework theories are rightly regarded as a great synopsis of many specific details. Nonetheless, here we have something that substantially differs, for example, from the unification of the electrodynamic theories (4.124)–(4.128) and, certainly, also from the possible unification of the four elementary interactions into one unified theory as currently sought. We will not attempt here to understand the difference in question in more detail, but only make the following remarks on it.

To make the many theories (e.g., for the numerous types of atoms and molecules) in quantum mechanics appear reduced to a single theory, i.e., general quantum mechanics, the Hamiltonian must be understood as a *variable*. For we had already pointed out in Chap. 2 that to consider a force field or of potential energy in classical mechanics as a constant of the theory means just to make these magnitudes external theory parameters, i.e., to assume that different force fields generate different theories. The same applies, of course, to the Hamiltonian of quantum mechanics: Only as a variable does it admit—ceteris paribus—to consider quantum mechanics as *one* theory in which the interaction is left just as open as the shape of the particular Ψ function. On the other hand, however, we have already indicated that general quantum mechanics (or, correspondingly, classical mechanics), understood in this way, raises the problem of whether it is still a well-interpreted physical theory at all. For, in order to interpret it, we would have to specify for each variable the domain of the elements indicated by it or, to put it more plainly, 'run through' by the variable. And, here, the question arises whether we can 'specify' such a domain, even in a vague sense, for something like the Hamiltonian. Certainly, in electrodynamics as a field theory, we are also faced with an immense abundance of possibilities. But these are, after all, only contingent expressions of one and the same interaction, whereas in the mechanical theories, in general, the possibilities in question are largely independent of each other.

We also have to point out a second danger in interpreting very general framework theories. Already for the previously mentioned theories, it repeatedly happens that two different Hamiltonians, for example, are not at all used to express different physical cases but rather *the same* one—only in such a way that it is treated once just in a certain approximation, but the other time exactly—or so it seems. Here, one replaces a Hamiltonian with another, not in order to treat a new physical case, but just the old one, yet simplified and, therefore, only approximately. This procedure, however, violates the normal conception of interpreting a variable, which in a certain domain, e.g., of possible Hamiltonians, can assume *any* value and with different values is also supposed to treat different cases, i.e., different contents or objects. How are we to exclude from this general rule those cases in which we want to use one value, e.g., one Hamiltonian, only as an approximation for another—i.e., for the treating the same case—only approximately? If such a case exists and we ask whether the approximating function may also play an independent role in another context, the question will probably be answered in the negative. But how do we know when we are dealing with one case and when with the other?

In this context, it should also be recalled that, in order to better understand the relationship between quantum mechanics and classical mechanics, new framework theories have been constructed that contain both theories—classical and quantum mechanics—as direct specialisations (cf. Chernoff and Marsden 1976 and further material in Chap. 4 of Vol. II). Here, then, what happens only incidentally in the previously mentioned cases is deliberately brought about: that different formal possibilities are used to treat the same case. Should we now say that classical and quantum mechanics have been reduced to the same theory (as a common generalisation)? That would be unhelpful, and we are not committed to do so unless it is made plausible that any model of such a common generalisation is physically possible. But that will not happen so easily. Nevertheless, such methods, which are also available with regard to general relativity (cf. Ehlers 1986), can provide certain insights that otherwise would be difficult to obtain.

References

Bogolubov, N. N., A. A. Logunov, A. I. Oksak, and I. T. Todorov. 1990. *General Principles of Quantum Field Theory*. Dordrecht: Kluwer.

Born, M. 1949. *Natural Philosophy of Cause and Chance*. Oxford: Oxford University Press.

Borsuk, K., and W. Szmieliew. 1960. *Foundations of Geometry*. Amsterdam: North-Holland.

Causey, R. L. 1977. *Unity of Science*. Dordrecht: Reidel.

Chernoff, P. R., and J. E. Marsden. 1976. Some Remarks on Hamiltonian Systems and Quantum Mechanics. In: W. Harper and C. A. Hooker (Eds.), *Foundations of Probability, Statistical Inference, and Statistical Theories of Science*. Dordrecht: Reidel, 35-53.

Earman, J. 1989. *World Enough and Space-Time*. Cambridge, MA: MIT Press.

Ehlers, J. 1986. On Limit Relations Between, and Approximate Explanations of, Physical Theories. In: R. B. Marcus et al. (Eds.), *Logic, Methodology and Philosophy of Science*, vol. 7. Amsterdam: North-Holland, 387–404.

Halmos, P. R. 1950. *Measure Theory*. Princeton, NJ: Van Nostrand.

Hempel, C. G. 1965. *Aspects of Scientific Explanation and Other Essays in the Philosophy of Science*. New York: The Free Press.

Hempel, C. G. 1966. *Philosophy of Natural Science*. Englewood Cliffs, NJ: Prentice-Hall.

Hoyningen-Huene, P. 1985. Zu Problemen des Reduktionismus der Biologie. *Philosophia naturalis* 22, 271–286.

Iyanaga, Sh., and Y. Kawada. 1977. *Encyclopedic Dictionary of Mathematics*. Two Vols. Cambridge, MA: MIT Press.

Ludwig, G. 1990. *Die Grundstrukturen einer physikalischen Theorie*. Berlin: Springer.

Medawar, P. 1974. A Geometric Model of Reduction and Emergence. In: J. Ayala and T. Dobzhansky (Eds.), *Studies in the Philosophy of Biology.* Berkeley: University of California Press, 57–63.
Nagel, E. 1961. *The Structure of Science.* New York: Harcourt, Brace & World.
Querenburg, B. V. 1973. *Mengentheoretische Topologie.* Springer: Berlin.
Rindler, W. 1960. *Special Relativity.* Edinburgh: Oliver & Boyd.
Schpolski, E. W. 1954. *Atomphysik,* Vol. 1. Berlin: Deutscher Verlag der Wissenschaften.
Scheibe, E. 1999. *Die Reduktion physikalischer Theorien. Ein Beitrag zur Einheit der Physik. Teil II: Inkommensurabilität und Grenzfallreduktion.* Berlin: Springer.
Shoenfield, J. R. 1967. *Mathematical Logic.* Reading, MA: Addison & Wesley.
Sikorski, R. [2]1969. *Boolean Algebras.* Berlin: Springer.
Sklar, L. 1967. Types of Inter-theoretic Reduction. *Brit. J. for the Phil. of Sci.* 18, 109–124.
Weizsäcker, C. F. V. 1980. *The Unity of Nature.* New York: Farrar, Straus and Giroux.
Weizsäcker, C.F. v., Scheibe, E., and G. Süßmann 1958. Komplementarität und Logik III. Mehrfache Quantelung. *Z. f. Naturforschung* 13a, 705–721.
Westenholz, G. V. 1978. *Differential Forms in Mathematical Physics.* Amsterdam: North-Holland.
Yoshida, R. M. 1977. *Reduction in the Physical Sciences.* Halifax, N. S.: Dalhousie University Press.

Chapter 5
Approximate Reductions

In all reductions of a theory T to a theory T' considered so far, the axioms of T were compatible with the axioms of T', and even T and T' themselves were compatible with each other. The second statement is not the same as the first, since, in addition to the theory axioms, the core axioms and the theory assertions (related to a specific domain of application) must also be taken into account. However, it must be asked what we mean here by 'compatibility'. The question "are Euclidean and hyperbolic geometry compatible?" does not allow a clear answer. If the concepts of point and distance between two points are given the same meaning in the two geometries—and this is probably what we are tacitly assuming—, then there is incompatibility. But there are interpretations of the hyperbolic within Euclidean geometry in which, especially, the two distance notions are interpreted in a different way; then the two geometries are compatible. More generally, if the formal languages of two theories overlap in their descriptive parts, this is in principle a focal point for incompatibility. But, if one then makes the overlap disappear simply by a renaming, the contradictions must in turn disappear. For two theories (each free of contradictions) with disjunct languages in the descriptive domain cannot contradict each other. Absolute signs are arbitrary, but relative ones are not. So, if we have two theories and are concerned with their relationship, we also have to specify the arguments of the formulae defining these theories in a way that is binding. There is then compatibility, depending on whether a common interpretation exists in *these* arguments or not.

As far as the exact reductions are concerned, we have every reason to check the compatibility since direct or indirect linguistic overlaps occur in all three respects mentioned at the beginning. For example, it is significant for the direct generalisation that in (4.4)–(4.6) of Sect. 4.1 all formulae Σ, Σ', and γ have the same arguments. Thus, Σ and Σ' could contradict each other in this constellation. Because of (4.4), however, this is impossible, provided Σ and Σ' are each separately free of contradiction. For, then, both axioms have certainly a common model. In the case of equivalences, where according to (4.46) of Sect. 4.2 the terms s in Σ and t in Σ' are connected by q, there is an indirect overlap. Again, however, it is easy to see

E. Scheibe, *The Reduction of Physical Theories*, Fundamental Theories of Physics 207,
https://doi.org/10.1007/978-3-662-65000-4_5

that structures $\langle X; s\rangle$ and $\langle X; t\rangle$ exist which together satisfy (4.46). The same can be immediately shown in all cases, even for the core axioms and theory assertions. The exact reductions by themselves can, therefore, not bring about any incompatibilities and also no genuine revisions, but this does not preclude their interpretation in terms of theoretical progress. Not to mention the possibility that theory unification can be also based on exact reductions (cf. Sect. 4.6), the improvements achieved by generalisation and refinement, as the main kinds of reduction, have already been discussed in detail in Chap. 4. In one sense, however, there is no improvement: Due to the above-mentioned compatibility, the new theory does *not* genuinely correct the overcome theory. Rather, exact reductions reproduce the most important aspects of the respective previous theory.

The most frequent type of theoretical progress in physics, however, always brings about a *correction in content* of the older theory: the new theory is incompatible with its predecessor, and this discrepancy can be more or less profound. For example, on a phenomenal level, the real gas equation has been only a small and easily comprehensible advance over the ideal gas equation, albeit one based on genuine incompatibility. In contrast, the step from classical mechanics to quantum mechanics involved profound changes of the older theory. Depending on whether the transformation is more or less harmless, it is less or more difficult to reconstruct the process *as a reduction*. After all, our maxim for a reduction was that the overcome theory should become redundant through its successor. But how is this possible if the two theories contradict each other—not to mention any deeper discrepancies? How can the new theory share the empirical successes of the older one (and thus be a candidate for improvement) if the latter contradicts the former?

The decisive idea for answering these questions is to admit *approximate reductions* in addition to the exact ones—reductions which do not permit the restoration of the older theory exactly, but only in a certain approximation. The question is, however, what it should mean at all to put a theory into an approximation relation to another theory. The word 'approximation' refers to topology, whereas theory relations, as shown in the previous chapter, are primarily established by means of logic. Should we, then, solve our problem by creating a new logic, a kind of topological logic, which has as its central basic concept that of an approximate implication (instead of an exact one)? Should we speak of one proposition not following exactly from another but only approximately? Without denying that this would be viable, let's take a different, more conservative route that lets logic be what it has always been.

As was already pointed out in Sects. 2.3 and 2.4, for us, a theory is first and foremost a concept. Therefore, a class of models corresponds to it as the scope of this concept. Since the theory has a core which includes all principal base sets, the class of models is an ordinary set typified by the principal base sets. For our standard example of a particle in the force field (Sect. 2.1), this theoretical scope or domain of the theory (as we called it there) is the set of solutions to the equation of motion for the particle (including m and α). Thus, if we have two theories with respective cores, we have two sets as their theoretical domains; and two sets can be compared topologically if they are subsets of one and the same topological space. In particular, the two sets can lie with respect to each other in such a way that they have common clustering points, or that one set belongs to the topological boundary of the other, or whatever else of this kind may happen. This would then give us the basis for an approximate comparison of the theories.

The approximate reductions to be treated in this chapter share certain, so to speak, pre-approximate features, which we want to put together already here. They are at the same time presuppositions for the approximation business to get under way at all. It was already said that we would welcome a common superior space for the two theoretical domains of the theories. And, still before any topological operation, we have to establish the already mentioned contradictory situation between the theories. So, in order to guarantee these matters in a not too complicated way, let us assume for this chapter that the core of the reduced theory T is a *fragment* of the core of the reducing theory T'. The two cores then have the (abbreviated) form

$$\langle X^\circ, X'^\circ; s^\circ, s'^\circ\rangle \text{ for } T', \quad \langle X^\circ; s\rangle \text{ for } T\ , \tag{5.1}$$

and, for the associated axioms Σ'° and Σ°, we want to assume:

$$\Sigma'^\circ(X, X'; s, s') \equiv \Sigma^\circ(X; s) \wedge s' \in \sigma'^\circ(X, X') \wedge \alpha'^\circ(X, X'; s, s')\ . \tag{5.2}$$

Σ'° and Σ° must be strong enough to obtain the necessary topological structures as being definable over the core. Concerning the axioms of T and T', we assume (2.47) of Sect. 2.3, as agreed. Furthermore, let the variable s_1 be typified by Σ' in the same way, i.e., by the same ladder term as by Σ. That Σ and Σ' contradict each other would then be ensured by

$$\vdash \neg\exists s_1, s_1'.\Sigma(X^\circ; s^\circ, s_1) \wedge \Sigma'(X^\circ, X'^\circ; s^\circ, s'^\circ, s_1, s_1')\,.\,. \tag{5.3}$$

On the other hand, the contradiction should not be too extensive, and this we can achieve by the existence of partial models of the form:

$$\vdash \exists s_1, s_2, s_2'.\Sigma(X^\circ; s^\circ, s_1) \wedge \Sigma'(X^\circ, X'^\circ; s^\circ, s'^\circ, s_2, s_2')\,.\,. \tag{5.4}$$

Only a requirement like this guarantees that we have a common stage for the intended topological comparative considerations.

5.1 Asymptotic Reductions

After these preliminary explanations, we turn to a first kind of approximate reductions—the asymptotic ones—and start again with an example. We choose it from the domain of the gas laws (cf. Becker 1966, Ch. 1 and 13). As is well known, the ideal gas equation

$$pv = RT \tag{5.5}$$

can first be improved by taking into account the finite volume b° of the gas molecules, resulting in the equation:

$$p(v - b^\circ) = RT \quad (b^\circ > 0)\,. \tag{5.6}$$

In a second step, we can take into account the forces that the gas molecules exert on each other. According to van der Waals, this leads to the equation

$$(p + a^\circ/v^2)(v - b^\circ) = RT \quad (a^\circ, b^\circ > 0) \tag{5.7}$$

with another constant a°. Without explicitly entering the molecular level at the moment, with the Eqs. (5.5)–(5.7), we face three phenomenological laws which are successive improvements of each other. But in what sense? Let us first reply that Eq. (5.5) can be asymptotically reduced to Eq. (5.6), which in turn can be reduced to (5.7), and thus, as will be shown, (5.5) can also be *asymptotically reduced* to (5.7). (The critical domain, in which (5.7) also describes the phase transition (incorrectly), can be neglected in the following considerations concerning reduction.)

To clarify the relations, we imagine that the Eqs. (5.5)–(5.7) are each reconstructed as the laws of a simple theory in the sense of Sect. 3.2., the van der Waals constants a° and b° being also constants in our sense, in addition to the (numerically fixed) gas constant R. Fixing the respective theory's core in this way indicates that we restrict ourselves to the theory of an arbitrary but definite kind of gas. Furthermore, the reconstruction as simple theories immediately shows that all the preconditions of our introduction are satisfied. In particular, our theories contradict each other in the sense that the Eqs. (5.5)–(5.7) pairwise have no common solutions in p, v, T. Moreover, we have the case, characteristic of asymptotic reductions, that the variables of the theories pairwise are of the same type. Therefore, everything here will take place on the common stage

$$M^\circ = (\mathbb{R}^+)^3, \tag{5.8}$$

that is, the set of possible values $p, v, T > 0$. Of these, the theoretical domains

$$M_i = \{p, v, T > 0|(1i)\} \quad (i = a, b, c) \tag{5.9}$$

each form a subset

$$M_i \subseteq M^\circ \quad (i = a, b, c)\,, \tag{5.10}$$

and these are now to be studied in their mutual relations.

To do this, we must first introduce a topology on M°, and here it is important to note that this can hardly be the standard topology of $\mathbb{R}^3$. In order to prove that, e.g., the solutions of (5.6) under certain conditions nearly coincide with solutions of (5.5), the topology must be *the same* as the one by means of which we rate the measurement inaccuracies to be taken into account for the measurement of pressure, volume, and temperature. For our explanation of (5.5) by (5.6) is also supposed to be empirical in the sense that, under certain conditions, the solutions of (5.6) agree with those of (5.5), not actually, but *within the measurement inaccuracy*. And, then, it would be intolerable if the topology in question would differ from the one used for expressing the measurement inaccuracy. Indeed, we have already introduced a uniform topology for ordinary magnitudes in Sect. 3.2, which then exclusively served to assess the measurement inaccuracies and was not the standard topology. Now, we take it also as a basis for the asymptotic reductions and others.

The uniform topology undoubtedly yet belongs to the vehicle of the reduction to be made. But an even more important element of the vehicle is a certain additional premise to the respective reducing law (5.6) or (5.7), which opens the way back to the law (5.5) or (5.6) to be reduced. In fact, the asymptotic reduction is an approximate version of the direct generalization of Sect. 4.1. And, as for the latter, the actual vehicle was the additional premise γ, we have something corresponding as well in the approximate counterpart: The solution sets M_i of (5.9) are, after all, simply surfaces in the 3-dimensional space M° of (5.8). A topological aspect of the reduction will be that these surfaces come arbitrarily close to each other, although they have no common points. The question is: Where do they do so? In the present case they come close at infinity. But even then we still have to specify a 'region' extending into infinity in M°, where the approximation takes place. This is exactly what the respective additional premise tells us. It is

$$b^\circ/v < \delta \tag{5.11}$$

for the reduction from (5.5) to (5.6), and

$$a^\circ/pv^2 < \delta\,. \tag{5.12}$$

for the reduction from (5.6) to (5.7). Finally, for the combined reduction of (5.5)–(5.7) we need the conjunction of the two premises (5.11)–(5.12). The new parameter δ already belongs to the topological statement which will define the asymptotic reduction in the following. In physical jargon, one would simply say so far: For volumes v which are large against b°, resp., for pressures and volumes for which pv^2 is large against a°, the solutions of (5.6), resp., (5.7) do not differ measurably from the solutions of (5.5), resp., (5.6).

For the reduction of (5.5) to (5.6), the precise topological statement that gives us the asymptotic reduction is:

For every $\epsilon > 0$ there is a $\delta > 0$, such that for each solution $\langle p, v, T\rangle \in M_b$ that *also* satisfies (5.11), there exists a solution $\langle p', v', T'\rangle \in M_a$ such that $|p' - p|/p,\ |v' - v|/v,\ |T' - T|/T < \epsilon$. (5.13)

The statement for the reduction of (5.7) to (5.6) is completely analogous. In the proof of (5.13) we have to satisfy the two existence statements by the choice

$$\delta = \epsilon,\ \text{as well as}\ v' = v - b^\circ,\ p' = p,\ T' = T\,.$$

The rest is mere calculation. It is obvious that, furthermore, the combination of the two reductions leads to another asymptotic reduction of (5.5) to (5.7) with the conjunction of the statements (5.11)–(5.12) as a new additional premise.

For further examples, we use the always rewarding theory of the *harmonic oscillator*. Here, reductions come into play by the fact that we have a classical and a quantum mechanical theory. Later, in Sect. 6.2, we will compare these two theories ab initio. For the moment, we are only concerned with the comparison of certain conclusions, which presuppose a very narrow class of (statistical) states, those in which we are dealing with an ensemble of oscillators of equal frequency ω in thermal equilibrium at temperature T. For the *mean energy* E, we then obtain classically

$$E = kT \ , \tag{5.14}$$

whereas, in the quantum mechanical case, it is

$$E = \hbar\omega/2 + \hbar\omega/\{\exp(\hbar\omega/kT) - 1\} \tag{5.15}$$

(cf. Cohen-Tannoudji et al. 1977, 620–623). So, classically, but not quantum mechanically, E is independent of ω. Again, in the 3-dimensional space of all $\omega, T, E \geq 0$, we are dealing with two surfaces (5.14) and (5.15), this time the asymptotic behavior being determined by the condition

$$\hbar\omega/kT < \delta \ . \tag{5.16}$$

As can be seen immediately from the transformation

$$E = kT\big[\tfrac{1}{2}\hbar\omega/kT + \hbar\omega/kT\big\{\exp(\hbar\omega/kT) - 1\big\}^{-1}\big] \ ,$$

for smaller and smaller δ in (5.16), the relation (5.15) turns into (5.14). This time the approximation takes place not only at infinity (high temperatures) but also along the straight line (5.14) in the plane $\omega = 0$.

This example also belongs to the simple theories in the terminological sense of Sect. 3.2, and the details in the sense of precise reduction statements (5.13) can be easily carried out. Still in the framework of simple theories, but on the way to a more complicated example, we consider further the relation between the *mean variances* σ^2 *of the spatial distributions* of a harmonic oscillator under the same physical conditions as before (cf. Cohen-Tannoudji et al. 1977, ibid.). Using the previous quantities and m as the mass of the oscillators, classically, we obtain

$$\sigma^2 = kT/m\omega^2 \tag{5.17}$$

and, quantum mechanically,

$$\sigma^2_{\text{qu}} = \sigma^2(\hbar\omega/2kT)\coth(\hbar\omega/2kT) \ . \tag{5.18}$$

Since, here, the factor at σ^2 approaches 1 for small delta of (5.16), we again face a case of the asymptotic reduction of a classical result to the corresponding quantum mechanical one. It shows, in addition, that the same applies to the spatial distributions themselves: For these are, respectively,

$$\mathrm{pr}(q) = \frac{1}{\sqrt{2\pi}\sigma} \exp\{-\tfrac{1}{2}q^2/\sigma^2\} \tag{5.19}$$

and

$$\mathrm{pr}_{\mathrm{qu}}(q) = \frac{1}{\sqrt{2\pi}\sigma_{\mathrm{qu}}} \exp\{-\tfrac{1}{2}q^2/\sigma_{\mathrm{qu}}^2\}\,. \tag{5.20}$$

This time, however, in addition to the real variables, T, ω, and m, a real function, pr, is involved. This calls for a few words about the topological situation.

Our stage M° is the product of the value sets for the three quantities T, ω, and m, and a suitable function space for the spatial distributions. In addition to the usual uniform topologies for the value sets, the uniform structure of uniform convergence is recommended for the function space (cf. Querenburg 1973, 160). For these, the $\langle f, f'\rangle$ with

$$|f(q) - f'(q)| < \epsilon$$

for all $q \in \mathbb{R}$ and any given ϵ form a fundamental system of environments. For the product space, this yields a canonical uniform structure (ibid., p. 121). Thus, then, a reduction statement of type (5.13) holds, completely analogous to the preceding cases. The least self-evident aspect of this procedure is the possibility of approximating the function $\mathrm{pr}_{\mathrm{qu}}$ in (5.18) *uniformly for all* $q \in \mathbb{R}$ by pr in (5.17). Here, this is the case. But, in general, it does not work that way, and we have to resort to local approximations (Sect. 5.3).

To obtain, now, also a general notion of asymptotic reductions, we proceed in several steps.

(1) Following up on premises (5.1)–(5.4) of the introduction, we additionally assume that Σ' contains no variables beyond those of Σ. Then, we have as the dramatis personae, first, the two theoretical domains:

$$\begin{aligned} M &= \{s \in \sigma(X^\circ) | \alpha(X^\circ; s^\circ, s)\} \\ M' &= \{s \in \sigma(X^\circ) | \alpha'(X^\circ, X^{\circ\prime}; s^\circ, s^{\circ\prime}, s)\} \end{aligned} \tag{5.21}$$

Here, as always, several s may be in play. To study the topological relation between M and M', we need a topology on a common superior space. This topology must be deducible as a uniform structure (cf. Sect. 3.2) from the core statement $\Sigma^{\circ\prime}$, i.e., with terms P° and q° internal with respect to $\Sigma^{\circ\prime}$ holds:

$$\begin{aligned} \Sigma^{\circ\prime}(X^\circ, X^{\circ\prime}; s^\circ, s^{\circ\prime}) \wedge M^\circ = P^\circ(X^\circ, X^{\circ\prime}; s^\circ, s^{\circ\prime}) \\ \wedge\, U^\circ = q^\circ(X^\circ, X^{\circ\prime}; s^\circ, s^{\circ\prime}) \vdash \mathrm{uni}(\langle M^\circ; U^\circ\rangle)\,. \end{aligned} \tag{5.22}$$

Finally, we need terms also internal for $\Sigma^{\circ\prime}$,

$$c_\delta(X^\circ; X^{\circ\prime}; s^\circ, s^{\circ\prime}; s)\ , \tag{5.23}$$

which we write for short c_δ, and which serves for the formulation of our respective asymptotic additional condition. c_δ carries the main part of the vehicle of an asymptotic reduction. As self-evident preconditions for everything else,

$$M, M', c_\delta \subseteq M^\circ \subseteq \sigma(X^\circ) \tag{5.24}$$

must be provable.

(2) We now come to some constraints on the sets derived in (1). First of all, there is the contradictory situation

$$M \cap M' = \emptyset \tag{5.25}$$

mentioned in the introduction. Further, it must be provable that the c_δ are a monotonically decreasing sequence of sets:

$$c_\delta \subset c_\epsilon \ \text{ for } \ \delta < \epsilon\ . \tag{5.26}$$

In this way, the additional constraint can be restricted more and more. However, it must be assured that

$$M' \cap c_\delta \neq \emptyset \tag{5.27}$$

for all $\delta > 0$, in order to obtain real 'conclusions' from these conjunctions.

(3) This brings us to the main condition, according to which, for a given accuracy, there are always regions in the model set of the reducing theory where all models agree, within the given measurement accuracy, with respectively one model of the theory to be reduced. To put it in precise terms:

$$\begin{aligned}&\text{For } u \in U^\circ \text{ there exists a } \delta > 0 \text{ such that } M' \cap c_\delta \subseteq M_u,\\ &\text{whereupon } M_u \text{ is a } u\text{-}neighborhood\ of\ M, \text{ that is,}\\ &y \in M_u \text{ if and only if a } y_1 \in M \text{ exists such that } \langle y, y_1\rangle \in u\ .\end{aligned} \tag{5.28}$$

This condition can be seen as a way to express an approximate conclusion conditioned by c_δ: it would be an exact conclusion if $M' \cap c_\delta \subseteq M$ held. But this is, of course, not possible because of (5.25) and (5.27). But, for sufficiently small δ, every model of M' can approximately be replaced by a model of M.

(4) The requirements imposed so far have, with (5.28), indeed arrived at a typical condition for an approximate reduction. But the peculiarity of an asymptotic reduction comes into view only if, in addition to (5.28) and (5.25),

$$C(M) \cap M' = M \cap C(M') = \emptyset \tag{5.29}$$

holds—$C(M)$ being the closure of M. Thus, from a topological point of view, M and M' are not necessarily completely separated. This would only be the case if, in addition to (5.29), even $C(M) \cap C(M') = \emptyset$ would hold. This would, of course, make any approximation impossible. Equation (5.29), however, still allows that M and M' have *common* clustering points (possibly at infinity), and this is the basis of (5.28). However, because of (5.29), these common clustering points are never also models of one of the two theories, and this distinguishes the asymptotic reduction from the limiting case reduction to be discussed in the following section. Related to this is the other condition for asymptotic reductions that, without prejudice to (5.25)–(5.27),

$$\bigcap_{\delta>0} M' \cap c_\delta = \bigcap_{\delta>0} M \cap c_\delta = \emptyset \tag{5.30}$$

holds, or sometimes even

$$\bigcap_{\delta>0} c_\delta = \emptyset\,. \tag{5.31}$$

Reductions and explanations appear especially convincing if they are strongly asymmetric. This is, e.g., met from the beginning for the micro-reductions to be considered later.[1] In contrast, asymmetry hardly seems to be met for the cases of asymptotic reduction considered so far. For, besides $M' \cap c_\delta \subseteq M_u$, one has there always also the converse statement $M \cap c_\delta \subseteq M'_u$ in (5.28) and, to that extent, symmetry. This symmetry is even significant for the demonstration of the empirical progress accompanying the reduction, as we will see later in this section. But we will also meet an example where this symmetry does not hold. Usually, however, one has at least the syntactic asymmetry that there is a constant in Σ', e.g., $\hbar$ in (5.15), which does not also appear in Σ, i.e., here in (5.14), and that this constant then also appears in c_δ, here in (5.16). Then (5.28) may indeed hold with the 'explanatory' statement $M \cap c_\delta \subseteq M'_u$. But it cannot contribute in this form to the explanation *by means of* Σ since c_δ is not an additional premise admissible for Σ.

After having clarified the general conditions, it will be useful to present in their light another, somewhat more complicated example of asymptotic reduction. Kepler's laws were the contemporary empirical basis for the corroboration of *Newton's theory of gravitation*. So Newton had to make sure that his theory 'provided' these laws in some sense. For more than two bodies, this is already a difficult mathematical problem, and a system such as that of the Sun, Earth, and Moon shows that Newton's theory goes far beyond Kepler's laws. In particular, it is not sufficient that the mass of the sun is large compared to that of the planets. These must also not come too close to each other. For only two bodies, however, one has the following simple consideration concerning the asymptotic reduction of Kepler's laws to Newton's theory (cf. Scheibe 1973b). Strictly speaking, this is a composite reduction, too, with the number of factors depending on how one puts the partner theories. We already mentioned in Sect. 4.2 that Kepler's laws contain a condition that rules out

[1] Chapter 3 of Vol. II.

the unconstrained orbits, which from Newton's point of view are completely on a par with the other solutions. So, here, *ahead* of everything else, there is a direct generalization without which Kepler's laws could not be explained. Likewise, after the reduction to Newton's equations, one could carry out a further reduction to general mechanics.

But the very core of the matter remains an approximate reduction, which, in the case of only two bodies, has to reduce the 'Kepler equations'

$$\begin{aligned} d^2\mathbf{x}'_1/dt^2 &= 0 \\ d^2\mathbf{x}'_2/dt^2 &= -m'_1(\mathbf{x}'_2 - \mathbf{x}'_1)|\mathbf{x}'_2 - \mathbf{x}'_1|^{-3} \end{aligned} \tag{5.32}$$

to the Newtonian equations

$$\begin{aligned} d^2\mathbf{x}_1/dt^2 &= -m_2(\mathbf{x}_1 - \mathbf{x}_2)|\mathbf{x}_1 - \mathbf{x}_2|^{-3} \\ d^2\mathbf{x}_2/dt^2 &= -m_1(\mathbf{x}_2 - \mathbf{x}_1)|\mathbf{x}_2 - \mathbf{x}_1|^{-3}, \end{aligned} \tag{5.33}$$

where the different denotation will be clarified below. Note that, in this form, both theories are invariant with respect to the Galilean group, and the following consideration does not violate this invariance in any regard. (In this respect, strictly speaking, we have a Galilei-invariant generalization of Kepler's laws, which are recovered only after adding the condition $\mathbf{x}_1 = 0$, i.e., with the sun at rest in the inertial frame. But, because this condition can always be achieved by an appropriate Galilean transformation, it is, on the other hand, not a physical restriction.)

The first thing to notice about the Eqs. (5.32)–(5.33) is that with the mass m_2 the Newtonian theory has a parameter not appearing in the Kepler equations. Since we aim at the moment at an asymptotic reduction, we consider m_2 as a constant m_2° and take as a common comparison space M° the totality of triples $\langle m_1, \mathbf{x}_1, \mathbf{x}_2\rangle$ with $m_1 > 0$ and sufficiently smooth functions $\mathbf{x}_i$. We leave the uniform structure of M° open for the moment. The solution sets M_a and M_b are defined by (5.32) and (5.33), respectively. To study them, we introduce the following transformation, M°.

$$\mathbf{x}'_1 = m_1\mathbf{x}_1 + m_2^\circ\mathbf{x}_2/(m_1 + m_2^\circ);\ \mathbf{x}'_2 = \mathbf{x}_2;\ m'_1 = m_1(1 + m_2^\circ/m_1)^{-2} \tag{5.34}$$

with inversion

$$\mathbf{x}_1 = \mathbf{x}'_1 - (m_2^\circ/f(m'_1, m_2^\circ))(\mathbf{x}'_2 - \mathbf{x}'_1);\ \mathbf{x}_2 = \mathbf{x}'_2;\ m_1 = f(m'_1, m_2^\circ). \tag{5.35}$$

We leave open the form of the function f. Its unique existence follows easily from (5.34) on the right. The transformation (5.34)–(5.35) has the property that it maps M_b one-to-one to M_a. For the solutions assigned to each other, we obtain further:

$$\begin{aligned} |m_1 - m'_1|/m_1 &= |1 - (1 + m_2^\circ/m_1)^{-2}| \\ |m'_1 - m_1|/m'_1 &= (1 + m_2^\circ/m_1)^2|1 - (1 + m_2^\circ/m_1)^{-2}| \end{aligned} \tag{5.36}$$

$$|\mathbf{x}_1 - \mathbf{x}_1'|/|\mathbf{x}_1 - \mathbf{x}_2| = m_2^\circ/m_1 + m_2^\circ; \quad |\mathbf{x}_2 - \mathbf{x}_2'|/|\mathbf{x}_1' - \mathbf{x}_2'| = 0\,. \tag{5.37}$$

Here, now, the following possibility opens: We provide M° with a uniform structure that is the product of the usual uniform structure

$$|m - m'|/m, |m' - m|/m' < \epsilon \tag{5.38}$$

for the masses and the uniform structure given by the fundamental system

$$|\mathbf{x}_1 - \mathbf{x}_1'|/|\mathbf{x}_1 - \mathbf{x}_2|; \quad |\mathbf{x}_2 - \mathbf{x}_2'|/|\mathbf{x}_1' - \mathbf{x}_2'| < \epsilon \tag{5.39}$$

for the *pairs* $\langle \mathbf{x}_1, \mathbf{x}_2 \rangle$. Thus, the latter is not once again the product of two uniform structures for the $\mathbf{x}_1$ and $\mathbf{x}_2$. Note, further, that in (5.39), first, a uniform structure is formed for the 6-dimensional configuration space of the *positions* of the two bodies and, only then (as a uniform structure of uniform convergence), for the totality of the *motions*.

We now see from (5.36)–(5.37) and (5.38)–(5.39) that we can get along with the condition of the mass m_1 being large compared to m_2, i.e.,

$$m_2^\circ/m_1 < \delta\,, \tag{5.40}$$

in order to claim the following.

> For a given $\epsilon > 0$, there exists a $\delta > 0$ such that for every
> Newtonian solution $\langle m_1, \mathbf{x}_1, \mathbf{x}_2 \rangle \in M_b$ with (5.41) there exists a
> Copernican solution $\langle m_1', \mathbf{x}_1', \mathbf{x}_2' \rangle \in M_a$ such that (5.38)–(5.39) holds.

For the proof, one simply chooses for $\langle m_1', \mathbf{x}_1', \mathbf{x}_2' \rangle$ the Kepler solution belonging to the given $\langle m_1, \mathbf{x}_1, \mathbf{x}_2 \rangle$ according to (5.34)–(5.35), as well as $\delta(2 + \delta) < \epsilon$. Note that, by the transformation (5.34)–(5.35), the 'sun' (as a body with the large mass m_1) is simply put into the center of gravity of the two-body system. From this, the 'true' (Newtonian) position of the sun may be miles away and, still, this distance is *small against* the (Newtonian) distance of the planet from the sun according to (5.39). If the bodies move on hyperbolas (in the center of gravity system), the distance of the sun from the center of gravity can become *arbitrarily* large and, nevertheless, it remains small against the distance to the planet.

Passing to questions of *empirical progress*, we must first draw attention to an oddity of asymptotic reductions. The idea that models of the reducing theory, if satisfying an additional condition, can be approximately replaced by models of the reduced theory underlies the technical formulation (5.28). The crucial relation which guarantees this substitution is the inclusion $M' \cap c_\delta \subseteq M_u$. If we are now concerned with proving the conservation principles C_3 and C_4 from Sect. 3.3; however, it soon becomes apparent that it is precisely the inversion $M \cap c_\delta \subseteq M_u'$ that we need for this purpose. Since we did not generally presuppose this, we now face the problem of *not* being able to prove C_3 andC_4 for asymptotic reductions in general. The inclusion of the inversion at issue in the notion of asymptotic reduction does not work as a way out since there are important cases for which it does not hold. One such case, which we will consider in more detail in Chap. 1 of Vol. II, is the well-known addition theorem of velocities: If (in the 1-dimensional case) two inertial systems move against each other with the relative velocity v, then, due to the Galilean transformation, the instantaneous 1-velocity u of an arbitrarily moving mass point transforms into u' according to

$$u' = u - v \,, \tag{5.41}$$

but, due to the Lorentz transformation, according to

$$u' = (u - v)/(1 - uv/c^2) \;\; (|u|, |u'|, |v| < c) \,. \tag{5.42}$$

The Galilean relation (5.41) can then be reduced to Einstein's (5.42) in terms of (5.28), with the *only* additional premise

$$|v| \ll c \,. \tag{5.43}$$

Just this reduction is mentioned in all textbooks of special relativity (see, e.g., Møller 1972, 49) and is, therefore, a classical example of an asymptotic reduction. But the inversion in question does not hold in its case: With the additional premise (5.43) alone, it is impossible to replace any model (5.41) by a model (5.42). After all, the velocities in (5.41) are not subject to any upper bound, and one need only choose $u' > c$ to see that this is impossible. If one adds to (5.43) also constraints on u and u', however, one obtains symmetrical relations also here. (Cf. the symmetry in the conditions D in Ehlers 1986.)

Thus, for the proofs needed to carry out the program indicated by (3.56) of Sect. 3.3, we assume that, besides $M' \cap c_\delta \subseteq M_u$, the inverse $M \cap c_\delta \subseteq M'_u$ is also available for the asymptotic reduction, and we do neither aim at full generality otherwise. As a *first main case*, we consider the case in which the structural elements of both theories may themselves be empirical data—1st order data, as we said in Sect. 3.1. Paradigmatic for this are the gas examples discussed at the beginning, in which the values of pressure, volume and temperature that characterize a system are also possible measuring data. For the reduced theory, we have then in general exact observation statements

$$s_1 = a_1, \ldots, s_m = a_m \tag{5.44}$$

with the measuring data a_i and the typifications

$$s_i \in \sigma_i(X^\circ) \,. \tag{5.45}$$

To these correspond approximate observation statements

$$\langle a_1, s_1 \rangle \in u_1, \ldots, \langle a_m, s_m \rangle \in u_m \tag{5.46}$$

with the measurement inaccuracies u_i, which originate from uniform spaces $\langle M^\circ; U^\circ\rangle$ with

$$u_i \in U_i^\circ, \quad M_i^\circ \subseteq \sigma_i(X^\circ)\,. \tag{5.47}$$

The space $\langle M^\circ; U^\circ\rangle$ responsible for the asymptotic reduction is the *uniform product* of the spaces $\langle M_i^\circ; U_i^\circ\rangle$ responsible for the approximation of the data (for the concept of product, see Querenburg 1973, 121). The connection is essentially given by the equivalence

$$\langle\langle a_1 \ldots a_m\rangle, \langle s_1 \ldots s_m\rangle\rangle \in u \leftrightarrow \langle a_1, s_1\rangle \in u_1 \wedge \cdots \wedge \langle a_m, s_m\rangle \in u_m\,, \tag{5.48}$$

where the u defined in this way form a fundamental system of neighborhoods in $\langle M^\circ; U^\circ\rangle$, if the same is true for the u_i with respect to $\langle M_i^\circ; U_i^\circ\rangle$.

We now look at the *conservation of compatibility* according to C_3 from Sect. 3.3. As is immediately clear from the gas examples, a measurement result compatible with M can be incompatible with M' if the measurement accuracy is too high, unless additional precautions are taken. Thus, the approximate compatibility

$$\exists s.\langle a, s\rangle \in u \wedge s \in M.\,, \tag{5.49}$$

with a for $\langle a_1, s_1\rangle$, s for $\langle s_1 \ldots s_m\rangle$ and u according to (5.48), is generally too weak a premise. But u is sufficiently imprecise if s is already in the asymptotic domain c_δ for which

$$M \cap c_\delta \subseteq M'_u \tag{5.50}$$

holds in the sense of (5.28), so that one can always choose δ for a given u such that (5.50) holds. Hence, instead of (5.49), we assume

$$\exists s.\langle a, s\rangle \in u \wedge s \in M \cap c_\delta.\,. \tag{5.51}$$

Then, because of (5.50), we obtain

$$\exists s.\langle a, s\rangle \in u \wedge s \in M'_u.$$

and, further,

$$\exists s.\langle a, s\rangle \in u \wedge \exists s'.\langle s, s'\rangle \in u \wedge s' \in M'..$$
$$\exists s'.\exists s.\langle a, s\rangle \in u \wedge \langle s, s'\rangle \in u. \wedge s' \in M'.$$

and, finally,

$$\exists s'.\langle a, s'\rangle \in u^2 \wedge s' \in M'.\,, \tag{5.52}$$

i.e., the approximate compatibility of a with M' with respect to u^2 (for the definition of the square, see Querenburg 1973, 112). So, except for the precaution with c_δ, we have to switch from the measurement inaccuracy u to u^2. But such incidentals are to be expected if we have only approximate relations.

For the treatment of the *conservation of success*, from now on, we will restrict ourselves to deterministic theories and observation statements with initial conditions. Instead of a, we abbreviate $\langle a_1 \dots a_m \rangle$ by $\langle A, a \rangle$, where A stands for $\langle a_1 \dots a_m \rangle$ and a for a_m. Then, we obtain exact observables of the form

$$S = A, \quad s = a\,, \tag{5.53}$$

as well as approximate ones of the form

$$\langle A, S \rangle \in U, \quad \langle a, s \rangle \in u\,, \tag{5.54}$$

where (5.47) is taken into account. We assume the determinism of the theory with the theoretical domain M to be of the form

$$\forall S. \exists! s. \langle S, s \rangle \in M..\,, \tag{5.55}$$

and, in this form, it holds (after appropriate restriction of S) also for any subset of M, e.g., for $M \cap c_\delta$. For our paradigm of the first main case, the gas example, S would stand for $\langle p, v \rangle$ and s for T. This determinism of the gas equation is, thus, based on the equation being uniquely solvable according to one or the other variable. Under the condition (5.55), the condition

$$\forall Ss. \langle S, s \rangle \in M \wedge S = A \to s = a. \tag{5.56}$$

for (5.53) to be a success for M turns out to be equivalent to

$$\langle A, a \rangle \in M\,. \tag{5.57}$$

But, then, the condition

$$\exists A'a'. \langle AA' \rangle \in U \wedge \langle aa' \rangle \in u \wedge \forall Ss. \langle Ss \rangle \in M \wedge S = A' \to s = a'.. \tag{5.58}$$

for approximate success (cf. (3.52) in Sect. 3.2) is equivalent to

$$\exists A'a'. \langle AA' \rangle \in U \wedge \langle aa' \rangle \in u \wedge \langle A'a' \rangle \in M. \tag{5.59}$$

and, therefore, to the approximate compatibility (5.49). Assuming that the reducing theory M' is also deterministic in the sense of (5.55), we can now conclude: Assuming (5.58) with $M \cap c_\delta$ in place of M, taking into account (5.50) with respect to $\langle U, u \rangle$, we obtain the compatibility (5.59) with $M \cap c_\delta$ in place of M and, from this, because of (5.50)–(5.52), correspondingly,

$$\exists S's'. \langle \langle Aa \rangle, \langle S's' \rangle \rangle \in V^2 \wedge \langle S's' \rangle \in M'. \tag{5.60}$$

with

$$\langle\langle Aa\rangle, \langle Ss\rangle\rangle \in V \equiv \langle AS\rangle \in U \wedge \langle as\rangle \in u \; . \tag{5.61}$$

Because determinism also holds for M', however, (5.60) now yields inversely:

$$\exists S's'.\langle\langle Aa\rangle, \langle S's'\rangle\rangle \in V^2 \wedge \forall Ss.\langle Ss\rangle \in M' \wedge S = S' \rightarrow s = s'.. . \tag{5.62}$$

That is, (5.54) is a success also for M', although with the increased inaccuracy V^2 defined via (5.61).

Let us now consider the *second main case*, in which not the s themselves but only their elements can be direct measuring data—the 2nd-order case in the sense of Sect. 3.1. For this, the particle theory of Sect. 2.1 is paradigmatic, with the motion of the particle as the structural element s at issue (and with fixed m and α). If we restrict ourselves also in the general case to *one* such s, then we have as exact observation statements

$$\langle a_1 \ldots a_m\rangle \in s, \;\; \langle b_1 \ldots b_m\rangle \in s, \ldots \tag{5.63}$$

and, therefore, as approximate statements (again with usual abbreviations):

$$\exists a'.\langle aa'\rangle \in v \wedge a' \in s., \;\; \exists b'.\langle bb'\rangle \in v \wedge b' \in s. \; \ldots \tag{5.64}$$

Whereas, in (5.44)–(5.48), we were dealing with observation statements with several s but always only one statement about them, we are now dealing with only *one* s (for simplification) about which, however, several observation statements are made. Both cases correspond to the fact that, altogether, we are dealing with statements about only *one* physical system. The typification is

$$s \in \mathrm{Pow}\{\theta_1(X^\circ) \times \cdots \times \theta_m(X^\circ)\} \; , \tag{5.65}$$

and, therefore, we need this time, besides the space

$$\langle N^\circ, V^\circ\rangle, N^\circ \subseteq \theta_1(X^\circ) \times \cdots \times \theta_m(X^\circ), \;\; v \in V \; , \tag{5.66}$$

which takes care of the approximations (5.64), another space

$$\langle M^\circ, U^\circ\rangle \text{ with } M^\circ \in \mathrm{Pow}(N^\circ) \tag{5.67}$$

on the power set for the asymptotic reduction; this space is, so to speak, the *uniform power space* of the previous one. It turns out that here, in general, the following structure has to be considered (cf. Bourbaki 1965, 227, Exercise 5): If $v \in V^\circ$ runs through its domain of definition, then the pairs

$$V = \{\langle x, y\rangle \in (M^\circ)^2 | x \subseteq y_v \wedge y \subseteq x_v\} \tag{5.68}$$

form a fundamental system of neighborhoods for a uniform structure on M°, where we assume, as usual,

$$p \in y_v \equiv \exists p' \in N^\circ.\langle pp' \rangle \in v \wedge p' \in y.. \tag{5.69}$$

Let us first also try to *conserve the compatibility* for the second main case,

$$\exists s.a \in s_v \wedge b \in s_v \wedge \cdots \wedge s \in M \cap c_\delta., \tag{5.70}$$

again taking into account c_δ, in such a way that corresponding to (5.50)

$$M \cap c_\delta \subseteq M'_V \tag{5.71}$$

holds and V is connected to v via (5.68). If we assume asymptotic reduction, this choice of δ is possible for any V, thus, also for any v. Now follows another calculation, starting from (5.61):

$$\exists s.a \in s_v \wedge b \in s_v \wedge \cdots \wedge s \in M'_V.$$
$$\exists s.a \in s_v \wedge b \in s_v \wedge \cdots \wedge \exists s'.\langle ss' \rangle \in V \wedge s' \in M'..$$
$$\exists s'.\exists s.\langle ss' \rangle \in V \wedge a \in s_v \wedge b \in s_v \wedge \cdots \wedge s' \in M'...$$

Now the term $\langle ss' \rangle \in V$ is distributed among the following terms.

$$\exists s'.\exists s.(a \in s_v \wedge \langle ss' \rangle \in V) \wedge (b \in s_v \wedge \langle ss' \rangle \in V) \wedge \cdots \wedge s' \in M'...$$

If we now calculate the terms in the parentheses, we get, e.g., for the first one

$$\exists a.\langle aa' \rangle \in v \wedge a' \in s. \wedge s \subseteq s'_v \wedge s' \subseteq s_v ,$$

with the consequence that, then, also:

$$\exists a'.\langle aa' \rangle \in v \wedge a' \in s'_v.$$
$$\exists a'.\langle aa' \rangle \in v \wedge \exists a''.\langle a'a'' \rangle \in v \wedge a'' \in s'..$$
$$\exists a'' \exists a'.\langle aa' \rangle \in v \wedge \langle a'a'' \rangle \in v. \wedge a'' \in s'..$$
$$\exists a''.\langle aa'' \rangle \in v^2 \wedge a'' \in s'..$$

Taking together these consequences, we finally obtain

$$\exists s'.a \in v^2(s') \wedge b \in v^2(s') \wedge \cdots \wedge s' \in M'., \tag{5.72}$$

where we have written $u(s)$ for s_u. Equation (5.72), however, means that the data are compatible with M' within the measurement inaccuracy v^2. Thus, we obtain exactly the same result as in the first main case.

Finally, as for the *conservation of success* in the second main case, we can prove it with similar ease by assuming, as in the first main case, that our theories are deterministic and that the observation statements contain associated initial conditions. The determinism, say for M, is then of the form

$$\exists! s.s \in M \wedge a' \in s \wedge b' \in s \wedge \cdots \wedge c' \in s.\,, \tag{5.73}$$

where the a', b', ..., e', as initial data, have to satisfy certain conditions, which we want to assume tacitly to be met in the following. Under this condition, a success of M is of the form

$$\begin{aligned} &\exists a'b' \ldots c'.\langle aa'\rangle \in v \wedge \langle bb'\rangle \in v \wedge \cdots \wedge \langle cc'\rangle \in v \\ &\forall s.s \in M \wedge a' \in s \wedge b' \in s \wedge \cdots \rightarrow c' \in s..\,, \end{aligned} \tag{5.74}$$

but the compatibility is of the form:

$$\begin{aligned} &\exists a'b' \ldots c'.\langle aa'\rangle \in v \wedge \langle bb'\rangle \in v \wedge \cdots \wedge \langle cc'\rangle \in v \\ &\wedge \exists s.s \in M \wedge a' \in s \wedge b' \in s \wedge \cdots \wedge c' \in s... \end{aligned} \tag{5.75}$$

Under the condition (5.73), the second line of (5.74) is equivalent to the second line of (5.65). As in the first main case, therefore, the proof for the conservation of success can be traced back to the proof just given for the conservation of compatibility, cf., in particular, (5.70)–(5.72).

5.2 Limiting Case Reductions

What happens if we integrate the van der Waals constants of the gas laws (5.5)–(5.7) into the theory, thus making them variables? Despite their usual designation as 'constants', there is of course nothing to preclude this step. With even greater right, we may ask what happens to the comparison (5.32)–(5.43) of Newton's theory of gravitation with Kepler's laws if we regard m_2° is no longer as a constant but as a variable. In Newton's theory, m_2 stands pari passu beside m_1, and their separate treatment (m_1 as a variable, m_2 as a constant) was done exclusively in favour of comparing it with Kepler's laws, where m_2 is missing. So what happens here, too, if we depart from the situation typical of asymptotic reduction, where both theories employ variables of equal number and significance, and move on to the case where the reducing theory has one or more *additional variables* compared to the reduced theory? This question, harmless as it seems, leads us to new kinds of approximate reductions. In fact, we have not yet fully exhausted the generality of approach (5.3)–(5.4) of the introduction to this chapter. There, on the one hand, the 'languages' of the two reduction partners were restricted in such a way that the language of the reduced theory is contained in that of the reducing one. On the other hand, we have already used the possibly greater linguistic resources of the reducing theory for the cores of the theory, but not yet for the variable parts.

By doing this now, we notice as a first difficulty that the theoretical domains of the two theories now no longer have the same typification; therefore, we have no natural common topological 'stage' for the comparison, as in the case of the asymptotic reduction. In the previous section, for the gas laws, this stage was clearly the appropriately topologised 3-dimensional space of all $p, v, T > 0$. But now, in the same case, in addition to this space, we have the other, 4-dimensional space of all $b, p, v, T > 0$ for the theory to be reduced, or even the 5-dimensional space of all $a, b, p, v, T > 0$ for the reducing theory. One law unfolds in one space, the other in the other. So where should we make the topological comparison and, thus, make a comparison at all? An obvious idea is to try it with both spaces and to adjust the respective other, 'dissimilar' law to the law distinguished by its topological space. However, it is then clear from the outset that we are dealing with a compound reduction, adding to the respective approximation an exact reduction that achieves this adjustment.

Let us first clarify the matter in more detail using the gas laws (5.5)–(5.7). It turns out that, under the new conditions, we can also carry out the reduction from (5.5) to (5.7) of the previous section, decomposing it into a reduction from (5.5) to (5.6) and one from (5.6) to (5.7). What really matters can, therefore, be explained by the reduction of

$$pv = RT \quad (p, v, T > 0) \tag{5.76}$$

to

$$p(v - b) = RT \quad (b, p, v, T > 0)\ , \tag{5.77}$$

where b is now a variable just as p, v, T (cf. Becker 1966, 66, for illustrating the values of b for different gases). Let us now proceed along the two proposed paths, first of all, from our usual point of view with regard to the topological-geometrical relations.

On the *first path*, let the stage space be given by the somewhat extended constraint in (5.77):

$$M^\circ = \{\langle b, p, v, T\rangle | b, p, v, T \geq 0\}\ . \tag{5.78}$$

Our law (5.76) is a hypersurface $M'' \subseteq M^\circ$ in it. If we want to compare (5.76) with (5.77), as a first step, an extension in the sense of Sect. 4.5 is recommended, adding the variable b, the constant 0 as a special case of b, and the associated space B° of values. Then, (5.76) is first reduced to the case

$$pv = RT, \quad b = 0 \quad (p, v, T > 0)\ , \tag{5.79}$$

and our theoretical domain is no longer M as the totality of p, v, T with (5.76), but M' as the totality of all b, p, v, T with (5.79). And, for this, M', now $M' \subseteq M^\circ$, holds in addition to $M'' \subseteq M^\circ$. This reduction is meaningful if one considers that the possibility $b = 0$ is indeed ruled out by (5.77) (finite volume of the molecules!), but (5.76) is just supposed to result as the limiting case for $b \to 0$. If we put $b = 0$ in (5.77), we even get (5.76). Of course, this should not mislead us into considering (5.76) as an exact special case of (5.77). It is not, because the theory (5.77) significantly excludes the possibility $b = 0$. But, as a *limiting case* of (5.77), we obtain (5.76) in the second step if we look at how M' lies to M'' in M°: M' is the *boundary* of the hyperface M'' (not belonging to M''), as can be clearly seen if we keep T fixed. Then, for M', we obtain an ordinary curve in the $\langle p, v, 0\rangle$ plane as the boundary of the ordinary surface M'' (for fixed T). This boundary results from the partition

$$M'' = \bigcup_{b>0} M''_b \tag{5.80}$$

with the surfaces

$$M''_b = \{\langle b_1, p, v, T\rangle | p(v - b) = RT \wedge b_1 = b\} \tag{5.81}$$

as the boundary position

$$\lim_{b \to 0} M''_b = M' \tag{5.82}$$

in a sense to be specified in more detail. Note, in particular, that here, as on the second path, *each* solution of (5.79) (and thus indirectly (5.76)) is a clustering point of the solutions of (5.77). In the asymptotic reduction of (5.5) to (5.6) in the previous section, *no* solution of (5.5) was a clustering point of solutions of (5.6). And this will be the general situation.

Let us now take the *second* of the proposed paths. Then, we have as a 'small' comparison stage

$$M^\circ = \{\langle p, v, T\rangle | p, v, T > 0\} \tag{5.83}$$

and, of course, $M \subseteq M^\circ$. The main question is how to adjust (5.77) and, thus, M'' to (5.83) in this case. It seems obvious to replace (5.77) by the much weaker statement

$$\exists b > 0 . p(v - b) = RT . \ (p, v, T > 0) . \tag{5.84}$$

Indeed, if we declare M' to be the totality of $p, V, T > 0$ with (5.84), then $M' \subseteq M^\circ$ according to (5.83), and it turns out, moreover, that M is again the boundary of M' (not belonging to M'). This time, the path of approximation

$$\lim_{b \to 0} M'_b = M \tag{5.85}$$

is given by the partition

$$M' = \bigcup_{b>0} M'_b \tag{5.86}$$

with the surfaces:

$$M_b' = \{\langle p, v, T\rangle | p(v-b) = RT\} . \tag{5.87}$$

Of course, this does not yet complete the reduction to (5.77). Not as the first term, as on the previous path, but as the second term, an exact reduction of (5.84) to (5.77) follows, namely, a refinement in the sense of Sect. 4.4. One passes from the very weak information (5.84) concerning the relation between p, V, and T to the much stronger (5.77), in which the variable b is again explicitly included, as one can easily see. In particular, here, we see how (5.76) is by no means directly reduced to (5.77), but only approximately to (5.84), and only a further reduction results in (5.77).

The relationships in the gas laws recur in many other cases of theory improvements in physics. For example, Kepler's 3rd law

$$a^3/T^2 = (1/2\pi)^2 m_1 \tag{5.88}$$

(m_1 the mass of the Sun) is a limiting case of Newton's improvement

$$a^3/T^2 = (1/2\pi)^2 m_1 \{1 + m_2/m_1\}^{-2} \tag{5.89}$$

for $m_2 \to 0$, where this m_2 is the now relevant mass of the planet. In the theory of the harmonic oscillator in thermal equilibrium, for $T \to 0$, the quantum mechanical limiting case

$$E = \tfrac{1}{2}\hbar\omega \tag{5.90}$$

results from the general formula

$$E = \tfrac{1}{2}\hbar\omega + \hbar\omega\{\exp(\hbar\omega/kT) - 1\}^{-1} . \tag{5.91}$$

Note the difference to the situation in the previous section: There, the classical law (5.14) is by no means a classical limiting case of (5.15) in the present sense. Rather, this law arose asymptotically from (5.15) for $\hbar\omega \ll kT$. It is true that the *straight line* $E = kT$ in the plane $\langle 0, T, E\rangle$ is indeed a limiting case of (5.15) for $\omega \to 0$. But the classical law $E = kT$ is meant in the sense of being independent of ω and represents a plane in the space $\langle \omega, T, E\rangle$. And this plane has just that asymptotic behaviour relative to the surface (5.15). Example (5.90)–(5.91) shows a case where the reduced theory does not even formally result as a special case of the reducing theory for a value forbidden by it (here $T = 0$). The case where the limit parameter does not have to approach 0 but tend to ∞ is similar. In the quantum mechanics of the hydrogen atom, the relation

$$E_n = -\alpha m_2(1/n^2) \quad (\alpha = e^4/2\hbar^2) \tag{5.92}$$

between the energy spectrum and the mass of the electron holds if the nuclear motion is not taken into account. If it is taken into account, the result is

$$E_n = -\alpha m_2(1/n^2)(1 + m_2/m_1)^{-1} , \tag{5.93}$$

where the mass m_1 of the proton now also appears. For $m_1 \to \infty$, (5.92) is the limiting case of (5.93). Again, the possibility of mere insertion is missing, pretending a mere special case.

We can easily find other examples from physics where we are in a similar situation as with the gas example given first. However, the following distinction should not be neglected. In all the examples given so far, the convergence in (5.82) and (5.85) takes place pointwise, but not in all of them uniformly. In fact, non-uniform convergence seems to be common: Limiting case theories arise from degenerate limiting values of the parameters still in play (cf. Berry 1994). If we look, e.g., at example (5.88)–(5.89), we find the following situation: It is true that for any given m_2 and $m_2 \to 0$, the surfaces in (5.89) again converge point-wise against the surface (5.88). But, in the critical region of $m_1 = 0$, they no longer do so uniformly. This can be seen immediately by looking at the family of functions of m_1 appearing in the form

$$(2\pi)^2 a^3 / m_1 T^2 = \{1 + m_2/m_1\}^{-2}$$

of (5.89) on the right, parameterised by m_2. These functions describe the respective deviation from the constant value 1 of (4.88). This deviation is not uniform: No matter how small we choose m_2, for sufficiently small m_1 the corresponding function approaches 0, no matter how long it has been close to 1 for the larger m_1. For the gas example (5.76)–(5.77), we read the same off from the transformation

$$pv - RT = pb\ .$$

b may be as small as you like, if it is given; for large p, we always get arbitrarily large deviations from (5.76). The situation is different in example (5.90)–(5.91). Here, we can immediately read off from the transformation

$$E - \tfrac{1}{2}\hbar\omega = kT(\hbar\omega/kT)\{\exp(\hbar\omega/kT) - 1\}^{-1}$$

that, for fixed T the deviation on the right for large ω approaches 0, for small ω it approaches kT, and for the rest it always remains below kT. For $T \to 0$, this results in uniform convergence, which, however, as said, is the exception rather than the rule.

Let us now describe *in general* what we should do in each of the examples and have already elaborated for the gas example. For the two theories Σ and Σ'', of which we want to explain the former as a limiting case of the latter, pre-approximately, we do not presuppose more than generally assumed in the introduction to this chapter. Now, we are concerned with the *new variables b* present in Σ'' as opposed to Σ, to which correspond certain constants b°, such that both originate from certain principal base sets B°. The b° and B° can be new (all, or some of them) as well, but do not have to be. After these explanations, we will henceforth write (emphasising what is significant)

$$\Sigma(\ldots s) \tag{5.94}$$

for the theory to be reduced (the limiting case) and

$$\Sigma''(\ldots B^\circ \ldots b^\circ \ldots s, b) \tag{5.95}$$

for the reducing theory. In addition to

$$b^\circ, b \in B^\circ\ , \tag{5.96}$$

from Σ'', it should follow that

$$(b \neq b^\circ)\ . \tag{5.97}$$

This expresses that, here, b° is the 'critical' value. If, as in example (5.92)–(5.93), the 'critical' value is improper (there, $m_1 = \infty$), we want to reframe it in such a way that the value at issue is transformed into the finite, such as ∞ to 0 by the transformation $b = 1/m_1$ in example (5.92)–(5.93). In certain cases, such as (5.76)–(5.77) and (5.88)–(5.89) above, the substitution of b° for b in Σ'' makes sense (albeit that it is illicit) despite (5.97) and yields Σ. In such cases, it is this conclusion that (illicitly) expresses the contradiction between Σ'' and Σ. Finally, we expect of Σ that

$$\Sigma''(X^\circ; \ldots, b) \wedge \Sigma''(X^\circ; \ldots, b') \vdash b = b'\ . \tag{5.98}$$

holds. In the above applications (the only), B° is canonically isomorphic to $\mathbb{R}^+ \cup \{0\}$ (= the set of non-negative real numbers), and $b^\circ = 0$ holds.

We saw that the reduction from Σ to Σ'' can be done in *two ways*. We had already met this possibility in Sect. 4.5. On both paths, an intermediate theory Σ' is used. On the first path, Σ is first exactly reduced to Σ'_1 and then Σ'_1 is approximately reduced to Σ''. On the second path, which runs through Σ'_2, it is the other way round: first comes the approximate and then the exact reduction. The exact reduction of the first path is an extension of Σ (cf. Sect. 4.5) to

$$\Sigma'_1(\ldots B^\circ \ldots b^\circ \ldots s, b) \equiv \Sigma^*(\ldots B^\circ \ldots b^\circ \ldots s) \wedge b = b^\circ\ , \tag{5.99}$$

where Σ^* is obtained from Σ through an insignificant core extension. This extension artificially introduces the variables b into Σ and makes Σ an improper special case of Σ''. On the second path, we first reduce Σ approximately to:

$$\Sigma'_2(\ldots B^\circ \ldots b^\circ \ldots s) \equiv \exists\beta.\Sigma''(\ldots B^\circ \ldots b^\circ \ldots s, \beta)\ . \tag{5.100}$$

The formation of Σ'_2 makes immediately clear that the step from it to Σ'' is also an extension in which the bound variable β simply becomes the free variable b.

For the two *approximations*, of Σ'' by Σ'_1 and of Σ'_2 by Σ, we first have to fix the two topological stages. For the purpose of making corroboration statements (cf. Sect. 3.2), Σ'' and Σ now each come with a uniform structure $\langle M_i^\circ, U_i^\circ \rangle$ referring to the variables s and b, or s, respectively. It would be useful, as done in the previous section, to claim this uniform structure also for the comparison with Σ'_1 or Σ'_2. On the first path, the difficulty arises that $\langle M_i^\circ, U_i^\circ \rangle$, ceteris paribus, may only be explained for B° *without* b° (since b° in Σ'' is a forbidden value) and cannot be continued to all of B°. However, this should be done in order to compare it with Σ_1', for which b° is explicitly permitted as a limiting value. We meet this difficulty in all the examples considered so far. In all these cases, B° is isomorphic to the non-negative real axis with 0 as the value for b°. However, the canonical uniform structure for positive magnitude values (cf. Sect. 3.2) cannot be continued to the value 0. We can only help ourselves by noting that this uniform structure and the ordinary (topological) metric of the real axis induce the *same topology*. Since in the present context only the latter will be important, we can therefore replace this inappropriate structure by the ordinary metric, which of course allows the required continuation. In general, we would then have

$$M_1^\circ \subseteq \sigma(X^\circ) \times B^\circ, \quad M_2^\circ \subseteq \sigma(X^\circ) \tag{5.101}$$

as the starting point for the further comparison. (Note that here $\sigma(X^\circ)$ as well as B° denotes a Cartesian product, depending on how many irreducible s or b are present.)

In M_1°, we then have to relate the theoretical domains

$$\begin{aligned} M' &= \{\langle s, b\rangle | \Sigma_1'(\ldots B^\circ \ldots b^\circ \ldots s, b)\} \\ M'' &= \{\langle s, b\rangle | \Sigma''(\ldots B^\circ \ldots b^\circ \ldots s, b)\} \end{aligned} \tag{5.102}$$

to each other, whereas in M_2° we have to relate:

$$\begin{aligned} M &= \{s | \Sigma(\ldots s)\} \\ M' &= \{s | \Sigma_2'(\ldots B^\circ \ldots b^\circ \ldots s)\} . \end{aligned} \tag{5.103}$$

As we already know from (5.80)–(5.81) and (5.86)–(5.87), the actual vehicles of the two reductions are then in each case a partition,

$$\begin{aligned} M'' &= \bigcup_{b \in I} . M_b''., \quad M_b'' \cap M_{b'}'' = \emptyset \ \text{ for } b \neq b' \\ M_b'' &= \{\langle s, b'\rangle | \Sigma''(\ldots B^\circ \ldots b^\circ \ldots s, b') \wedge b' = b\} \end{aligned} \tag{5.104}$$

and

$$\begin{aligned} M' &= \bigcup_{b \in I} . M_b'., \quad M_b' \cap M_{b'}' = \emptyset \ \text{ for } b \neq b' \\ M_b' &= \{s | \Sigma''(\ldots B^\circ \ldots b^\circ \ldots s, b)\} , \end{aligned} \tag{5.105}$$

respectively, where I is the set defined by

$$I = \{b \in B^\circ | \exists s . \Sigma''(\ldots s, b)\} . \tag{5.106}$$

Nevertheless, even without vehicles, a significant topological statement about the relation of Σ_1' to Σ'' and of Σ to Σ_2' as a limiting case relation can be made.

We formulate it as the first of a total of three statements describing the *limiting case reduction* in general. It is the same for the pairs M', M'' and M, M'. Since we do not want to complicate the terminology, we formulate it for the second pair (= second path). We have to note the preconditions

$$M, M' \subseteq M_2^\circ, \quad M \cap M' = \emptyset\,, \tag{5.107}$$

and to specify a topology on M_2°. The first relation in which Σ and Σ_2' stand to each other is that M belongs to the *outer boundary* of M', i.e., that

$$M \subseteq \{C(M') \cap C(M_2^\circ \backslash M')\} \backslash M' \tag{5.108}$$

holds, where in general $C(M)$ is the closure of M. The expression in the curly brackets denotes the boundary of M'. We call the outer boundary those boundary points that do not belong to M'. The restriction to the outer boundary in (5.108) takes into account that our theories contradict each other, so (5.107) on the right holds. It would not suffice that M belongs to the boundary of M'. In case (5.102) for the gas laws (first path), all of M'' belongs to its (own) boundary in M_1° The condition that M' also belongs to it would be inappropriate. Furthermore, M need not be the whole outer boundary of M': In case (5.90)–(5.91), besides (5.90), the straight line $E = kT$ in the plane $\langle 0, T, E\rangle$ also belongs to the outer boundary of (5.91). Which part of the outer boundary of M' is occupied by M is determined by the decomposition (5.105). One easily proves that all previous examples satisfy (5.107)–(5.108) if we take the ordinary metric (based on the absolute value) for the real variables. Since (5.108) is a purely topological statement (as opposed to a 'uniform' statement) and the ordinary topology coincides with the topology of the uniform structure for quantity values (according to Sect. 3.2), the condition (5.109) conforms to our other specifications in these special cases.

Equation (5.108), however, is not a limit statement. It does not represent the model set M as a limit, although according to (5.108) every model from M is a clustering point of models from M'. But we have not yet generally justified statements (5.82) and (5.85), with which a limit statement should, indeed, be made for the gas example, with (5.107)–(5.108). The explicit presentation of a reduced theory as a limit is of interest to distinguish the reductions considered now from the asymptotic ones of the previous section, among other things. For the latter do not result in forming a limiting case. At best, we could say that the reduced theory forms an asymptote for the reducing one—but not a limiting case. A first possibility to take the expression of the limiting case at its word is offered in the present context by treating the theoretical domains M' and M'' (first path) or M and M' (second path) as *wholes*—not looking inside them, so to speak, to obtain the relevant relationship. This means concretely to introduce a topology or even a uniform structure on the power set of M_i° and then invoke the partitions (5.104)–(5.105).

If $\langle M_i^\circ, U_i^\circ \rangle$ is a uniform structure, then we easily obtain a uniform structure on $\mathrm{Pow}(M^\circ)$ in the following way: If $u \in U$ runs through all neighbourhoods of U°, then the resulting set

$$\{\langle M, N\rangle | M, N \in \mathrm{Pow}(M^\circ) \wedge M \subseteq N_u \wedge N \subseteq M_u\}_{u \in U^\circ} \tag{5.109}$$

forms a fundamental system of neighbourhoods of a uniform structure $U^{\circ\circ}$ on $\mathrm{Pow}(M^\circ)$ (cf. Bourbaki 1965, 227). In it, N_u is generally defined by

$$N_u = \{p \in M^\circ | \exists q.\langle p, q\rangle \in u \wedge q \in N\} \tag{5.110}$$

for given $N \in \mathrm{Pow}(M^\circ)$. Now, we consider the decompositions (5.104)–(5.105). In (5.105), for example, (again restricting ourselves to the second path) the family $\{M_b'\}_{b \in B^\circ}$ expresses the improvement over M, insofar as there the theory explicitly takes the variable b into account, whereas b does not appear at all in M, since M is simply the limiting case

$$M = \lim_{b \to b^\circ} M_b' \,. \tag{5.111}$$

But what does (5.111) mean? First of all, with M_b' we have a function (of b) that maps $B^\circ \backslash \{b^\circ\}$ into $\mathrm{Pow}(M_2^\circ)$, and since both spaces have an (even uniform) topology, it makes sense to demand that this function is *continuous*. Moreover, this requirement is reasonable with respect to (5.111). We might, then, state the limiting case condition in such a way that we require M_b' to have the value M for b° and to be continuous *also* in b°. In other words, we require this function to be *continuously continuable* from $B^\circ \backslash \{b^\circ\}$ to B°. Equivalent to this is the requirement that M_b' converge as a function of b for the image of the neighborhood filter of b° (in B°) in $\mathrm{Pow}(M_2^\circ)$ with value M. Under the same conditions as for (5.107)–(5.108), (5.109)–(5.111) is also satisfied for all the examples discussed above.

In the treatment of the examples, we also mentioned that the convergence against a limiting case is pointwise but, in general, not uniform. For a family of functions, it is clear what pointwise convergence is, since here we have the same domain of arguments for all functions. In (5.82) and (5.85), however, it was primarily not about functions but about hyperfaces in a number space. Even if these can be represented as functions in various ways, we had not distinguished such a representation. And, in the general case, it may not be possible at all. But what can be done (we restrict ourselves again to the second path (5.105)) is to assume a continuous function

$$f : M \times I \to M' \tag{5.112}$$

that connects M as part of the outer boundary of M' pointwise with the interior of M'. This is achieved in detail by requirements such as

$$f(p,b) \in M'_b(\text{or: } \Sigma''(\dots f(p,b),b))$$
$$p \neq p_1 \vdash \underset{b\in I}{\forall}.f(p,b) \neq f(p_1,b).\,, \tag{5.113}$$

but, above all, by the crucial limit condition

$$\lim_{b\to b^\circ} f(p,b) = p\,. \tag{5.114}$$

This would be the pointwise convergence (for one $p \in M$ respectively), and we have already seen that more (in terms of uniform convergence) cannot be defended in general.

The three conditions (5.107)–(5.108), (5.109)–(5.111), and (5.112)–(5.114) are not independent of each other. Thus, the limit statement (5.114) implies the partial statement $M \subseteq C(M')\backslash M'$ of (5.108). However, here we do not want to follow up on such dependencies. Instead, let us briefly look back at the asymptotic reduction and ask about its connection with the limiting case reduction of this section. It becomes most clear if we go back to the second path and look at the M'_b, in relation to M from (5.103, 5.105). It is significant for the limiting case reduction that, according to (5.111) strengthened by (5.114), the M'_b converge against M. In contrast, for the asymptotic reduction, we stated that the domain M is an asymptote for M'_b for arbitrary but *constant* b. This statement is obviously completely independent of the limiting case statement with variable b: In principle, all M'_b can have M as their asymptote without also converging against M for $b \to b^\circ$; conversely, this convergence can take place without a single M'_b having M as its asymptote. So we are really dealing with two completely different situations. However, in physics, it is usually the case that both situations—convergence and asymptotics—are found. Furthermore, the asymptotic situation seems to prevent uniform convergence.

Finally, let us again look at the progress associated with limiting case reduction! It turns out that, in order to prove the conservation postulates C_3 and C_4 of Sect. 3.3, we only need (5.108) of the claims (5.107)–(5.114) defining the limiting case reduction, and, of it, only the conclusion that every model of M is a clustering point of models of M'. It further turns out that, on general considerations, the proof in question looks essentially the same for the two paths of an approximation of Σ'_1 by Σ'', on the one hand, and of Σ by Σ'_2, on the other (cf. (5.99)–(5.100) and (5.101)–(5.106))—provided that for the first path the mentioned topological problem is solved. The proof of the conservation of compatibility for the first main case with the elements s as possible measurement data (cf. the previous section) would, thus, begin with the mentioned presupposition

$$\langle s\dots\rangle \in M \vdash \exists s'.\;\dots\langle s's\rangle \in u \wedge \dots \wedge \langle s'\dots\rangle \in M'. \tag{5.115}$$

that each model of M is a clustering point of models from M'. That the observation statements are approximately compatible with measurement data s° with M means

$$\exists s.\;\dots\langle ss^\circ\rangle \in u \wedge \dots \wedge \langle s\dots\rangle \in M. \tag{5.116}$$

with neighbourhoods u of the uniform structure on M°. If one puts the terms preceding $\langle s \ldots \rangle$ and the existence constraint in front of both sides of the implication (5.115), one obtains

$$\exists s. \ldots \langle ss^\circ \rangle \in u \wedge \ldots \exists s'. \ldots \langle s's \rangle \in u \wedge \ldots \langle s \ldots \rangle \in M'..$$

as a consequence of (5.116). Exchanging the existential quantifiers then yields terms of the form

$$\exists s. \langle s's \rangle \in u \wedge \langle ss^\circ \rangle \in u.$$

and one immediately obtains:

$$\exists s'. \ldots \langle s's^\circ \rangle \in u^2 \wedge \cdots \wedge \langle s' \ldots \rangle \in M'.. \quad (5.117)$$

But this is the compatibility of the data $s^\circ, \ldots$ with M', where again the original measurement inaccuracies $u, \ldots$ are replaced by their squares $u^2, \ldots$.

In the previous section, we had shown that Kepler's (simulated) two-body theory is asymptotically reducible to the corresponding Newtonian theory with the asymptotic constraint (5.40) that the mass m_1 of one body is large against the mass m_2 of the other. $m_1 = m_2^\circ$ had been regarded as a constant there that only appeared in the Newtonian theory. Since for $m_2 = 0$ the Newtonian equations (5.33) turn into the Keplerian ones (5.32), it is obvious to assume that the latter are also in the strict sense a limiting case of the former for $m_2 \to 0$. This issue is in need of proof even if an asymptotic reduction has already been proven. For we had already emphasised that the two reductions are fundamentally independent of each other. Now there can be no doubt about the provability of this limiting case either (cf. Scheibe 1973b). However, we do not want to deal with it here, but only briefly hint at its implications for the conservation of compatibility. The following formulations for this special case give a good account of the general situation in which we are primarily interested.

Since we are dealing with a concrete case, we have to choose one of the two paths (5.99)–(5.100). We go the second path and then have to compare the Kepler theory with the theory

$$\langle \mathbf{x}_1, \mathbf{x}_2, m_1 \rangle \in \mathrm{New}_2' \equiv \exists m_2. \langle \mathbf{x}_1, \mathbf{x}_2, m_1, m_2 \rangle \in \mathrm{New}. \quad (5.118)$$

where we let New be defined by (5.33) of the previous section. To simplify the further formulae, we write f and g for the motion (space-time function) of the two bodies and m for the mass of one. That every Kepler solution according to (5.32) is a clustering point of a Newton solution in the sense of (5.36) now means:

$$\begin{aligned}\langle f, g, m \rangle \in \mathrm{Kep} \vdash \exists f', g', m'. \langle f' f \rangle \in V \wedge \langle g' g \rangle \in W \\ \wedge \langle m' m \rangle \in u \wedge \langle f' g' m' \rangle \in \mathrm{New}_2'..\end{aligned} \quad (5.119)$$

Here, as in (5.68) of the previous section,

$$\langle f'f\rangle \in V \equiv f' \subseteq v(f) \wedge f \subseteq v(f') \tag{5.120}$$

are the neighbourhoods in the uniform power of the uniform structure of v on the set of spacetime points (which we leave open here), and $v(f)$ is the v-neighborhood f according to (5.69). That given data are compatible with Kepler's theory now means

$$\begin{aligned} &\exists fgm.\langle x^\circ t^\circ\rangle \in v(f) \wedge \dots \wedge \langle \xi^\circ \tau^\circ\rangle \in w(g) \wedge \dots \\ &\quad \wedge \langle mm^\circ\rangle \in u \wedge \langle fgm\rangle \in \text{Kep.} . \end{aligned} \tag{5.121}$$

A mere calculation similar to that of (5.70)–(5.72) then makes use of (5.119)–(5.120) to infer from (5.121) that the data are also compatible with New_2', passing again to the squares of the neighbourhoods:

$$\begin{aligned} &\exists f'g'm'.\langle x^\circ t^\circ\rangle \in v^2(f) \wedge \dots \wedge \langle \xi^\circ \tau^\circ\rangle \in w^2(g) \wedge \dots \\ &\quad \wedge \langle m'm\rangle \in u^2 \wedge \langle f'g'm'\rangle \in \text{New}_2'. . \end{aligned} \tag{5.122}$$

5.3 Local Reductions

The concepts of approximate reduction developed in the previous two sections are not sufficient to cover even the most simple approximate reductions that occur in physics. We will soon see that, for example, the reduction of the Rayleigh-Jeans law of radiation to that of Planck and the reduction of the Galilean law of falling bodies to that of Newton do not belong to the kinds of reductions considered so far. For we have, hitherto, assumed that the sets M and M' which enter directly into the reduction statement are the sets of all solutions of a theory T and T', respectively, the solutions being taken as *wholes*. In the case of the laws of radiation and falling bodies, however, there is no approximate comparison referring directly to whole system descriptions. Rather, one must resort to *partial descriptions* in order to arrive at reasonable approximations. In the case of functions, such as the energy densities of the cavity radiation or the position-time functions of the laws of falling bodies, one uses partial descriptions that restrict these functions to intervals smaller than the 'natural' intervals of the independent variable. The approximations can then become very weak due to these intervals becoming arbitrarily small. For this reason, we speak here of *local reductions*. The local consideration mode is known, in particular, from differential geometry, and examples from this field also play a role in physics. From the outset, however, we have to emphasise that the point of view of localisation does not lead us to compare and reduce new 'local theories'. We are, rather, dealing with the reduction of the original theories, and only the reduction *itself* requires the local approach. Even the use of uniform structures, which are in a certain sense localisations of other uniform structures, does not indicate that we are dealing with local reductions. To make this absolutely clear, let us start with an asymptotic reduction that already employs a localised uniform structure.

The case is the theory of the plane pendulum, already used in Sect. 4.4 as an example for the refinements, with the equation of motion

$$d^2\varphi/dt^2 + \omega^2 \sin\varphi = 0\,, \tag{5.123}$$

which is often replaced for small amplitudes

$$|\alpha_\varphi| \ll \pi \tag{5.124}$$

by the oscillator equation

$$d^2\varphi/dt^2 + \omega^2\varphi = 0\,, \tag{5.125}$$

taking advantage of the fact that for small amplitudes $\sin\varphi \approx \varphi$. This looks like an asymptotic reduction, in which (after Sect. 5.1) M' is the solution set of (5.123), M is that of (5.125), and the sets c_δ constrain the functions from M' according to (5.124). However, even under the constraint (5.124), there is no question of approximating a given solution of (5.123) on the *whole* real time axis by a solution of (5.125) for this reduction. Even for small amplitudes, the differences $|\varphi'(t) - \varphi(t)|$ will always get to the order of magnitude of the amplitudes themselves, if one only waits long enough. For the solutions φ' of (5.123) and the solutions φ of (5.125) have different frequencies, which induce this 'build-up'. This circumstance can be taken into account by choosing a uniform structure on a suitable set M° of bounded, smooth functions, which accomplishes the approximations only on *finite* intervals. Such a uniform structure could be generated by the sets $u(\epsilon, I)$, defined by the totality of pairs $\langle\varphi', \varphi\rangle$ with

$$|\varphi'(t) - \varphi(t)|/\max\{\alpha_{\varphi'}, \alpha_\varphi\} < \epsilon \tag{5.126}$$

for all $t \in I$ with a finite interval I, for example. With this, then, the usual asymptotic reduction holds: For a neighbourhood u, a $\delta > 0$ exists such that for every φ' from (5.123) with an amplitude $\alpha' < \delta$ there exists a φ from (5.125) with $\langle\varphi', \varphi\rangle \in u$ (cf. Lin and Segel 1974, 2.2, 7.1, and 11.2-3). The necessary restriction to partial descriptions, here the restriction of the solutions to finite intervals, therefore, only concerns the choice of the uniform structure and does not affect the reduction statement itself.

Even for limiting case reductions (cf. the previous section), it may happen that only local approximations are possible. As an example, let us take the theory of two linear oscillators with the equations of motion

$$\begin{aligned} m(d^2x/dt^2) &= k(y-x) \\ \mu(d^2y/dt^2) &= -k(y-x) \end{aligned} \tag{5.127}$$

as an improvement of the theory

$$\begin{aligned} d^2x/dt^2 &= 0 \\ \mu(d^2y/dt^2) &= -ky \end{aligned} \tag{5.128}$$

in which the first particle is considered to be infinitely heavy (cf. (6.33)–(6.45) of Sect. 6.2, where the analogous quantum mechanical case is considered). If we consider the first particle as being at rest, as usual, and correspondingly the centre of gravity in (5.127), we obtain for (5.127) the solutions

$$\begin{aligned} x' &= -\mu/m_S(\alpha\cos\omega_R t + \beta\sin\omega_R t) \\ y' &= m/m_S(\alpha\cos\omega_R t + \beta\sin\omega_R t)\,, \end{aligned} \tag{5.129}$$

with

$$m_S = m + \mu,\quad m_R = m\mu/(m+\mu),\quad \omega_R^2 = k/m_R\,, \tag{5.130}$$

and for (5.128), correspondingly,

$$\begin{aligned} x &= 0 \\ y &= \gamma\cos\omega t + \delta\sin\omega t\,, \end{aligned} \tag{5.131}$$

with

$$\omega^2 = k/\mu\,. \tag{5.132}$$

The theory (5.131)–(5.132) is obviously a limiting case of (5.129)–(5.130) for $m \to \infty$, since for this limit $\mu/m_S \to 0$ and, therefore, $x' \to x$ even on the whole axis. The corresponding $y' \to y$, however, results again only from a uniform structure that corresponds to (5.126) and aims at finite intervals. The three aspects (5.107)–(5.114) of the limiting case reduction of the previous section can thus be exemplified. For example, we may choose the function f according to (5.112)–(5.114) here, in such a way that we start from a solution $\langle\mu, \gamma, \delta\rangle$ of (5.131) and assign to it, together with m, the solution y' of (5.129) with

$$\alpha = (m_S/m)\gamma,\quad \beta = (m_S/m)\delta\,.$$

Then, we obtain

$$|y(t)' - y(t)| = |\gamma(\cos\omega_R t\ - \cos\omega t) + \delta(\sin\omega_R t\ - \sin\omega t)|\,.$$

Because of $\omega_r \to \omega$ for $m \to \infty$ and (for the first summand)

$$\cos\omega_R t\ - \cos\omega t = 2\sin\left(\tfrac{1}{2}(\omega_R - \omega)t\right)\sin\left(\tfrac{1}{2}(\omega_R + \omega)t\right)\,,$$

we see that the above difference becomes arbitrarily small on finite intervals.

The following case, however, shows a completely different behaviour. In the theory of cavity radiation (cf. Schpolski 1954, Chap. 6), initially the law of Rayleigh and Jeans,

$$\rho(\nu) = (8\pi\nu^2/c^3)kT\,, \tag{5.133}$$

and that of Wien,

$$\rho(\nu) = (8\pi\nu^3/c^3)kT\exp\{-h\nu/kT\}\,, \tag{5.134}$$

were known. The former agreed well with the measurements in region

$$h\nu \ll kT\,, \tag{5.135}$$

the latter in region

$$h\nu \gg kT\,. \tag{5.136}$$

Only Planck was then able to establish his famous law

$$\rho(\nu) = (8\pi\nu^3/c^3)(\exp\{h\nu/kT\} - 1)^{-1} \tag{5.137}$$

in such a way that it resulted in empirically satisfactory values for the entire spectrum. In addition, it approximately explained the laws of Rayleigh-Jeans and Wien in the above domains (5.135) and (5.136), respectively. For the first case, this becomes evident by writing Planck's law in the form

$$\rho(\nu) = (8\pi\nu^2/c^3)kT(h\nu/kT)(\exp\{h\nu/kT\} - 1)^{-1}$$

and, for the second case, by neglecting the 1 in $\exp\{h\nu/kT - 1\}$. In addition, however, it becomes clear that this consideration made at fixed T by no means concerns the functions in question for the whole ν spectrum. The conditions (5.135)–(5.136) do not restrict a set of complete system descriptions to a subset, as in the case of the asymptotic reductions. Rather, they restrict the domain of definition of each single (here only: one) complete system description. And it is this localising step which then makes the approximation possible.

In order to arrive at a clear notion of local reduction, let us carry out the example a little further. First of all, the radiation laws (5.133)–(5.134) and (5.137) cannot be reconstructed as simple theories in the sense of Sect. 3.2, which connect the three magnitudes ρ (spectral energy density), ν (frequency), and T (temperature). Although this is mathematically possible (writing ρ instead of $\rho(\nu)$ in (5.133)–(5.134) and (5.137)), it would not correspond to the physical situation, which is characterised by a thermal equilibrium, i.e., by a mixture of waves of *all* frequencies. The system is, therefore, to be described by its temperature and the *function* ρ, which indicates the distribution of the energy density over all frequencies. And precisely these two parameters are related to each other by the laws (5.133)–(5.134) and (5.137). Of these, it is then the function ρ that forces a local approach. Equations (5.135)–(5.136) restricts the domain of definition of each of the functions ρ in $\langle\rho, T\rangle$ rather than the set of solutions $\langle\rho, T\rangle$ of (5.137). For example, we have the following approximate situation for (5.133) and (5.137) with respect to (5.135):

For every solution $\langle\rho, T\rangle$ of (5.137) exists a solution $\langle\rho', T'\rangle$ of (5.133), such that for every $\epsilon > 0$ there exists a $\delta > 0$, such that for all ν with

$$h\nu/kT < \delta$$

hold the inequalities:

$$|\rho'(\nu) - \rho(\nu)|/\rho'(\nu) < \epsilon \, , \quad |T - T'|/T' < \epsilon \, .$$

For a given solution $\langle \rho, T \rangle$ of (5.137), the solution $\langle \rho', T' \rangle$ of (5.133) is chosen such that $T' = T$ and ρ' is the solution of (5.133) uniquely determined by this T. For a given ϵ, δ is chosen such that

$$1 - \delta/(\exp \delta - 1) < \delta \, .$$

A simple calculation then gives

$$|\rho(\nu) - \rho'(\nu)|/\rho'(\nu) < \epsilon$$

for all ν with $h\nu/kT < \delta$.

As a second example, we mentioned Galileo's and Newton's laws of falling bodies (cf. Vollmer 1989). Restricting ourselves to the one-dimensional case, we want to accept Galileo's law in the (modern) form

$$d^2x/dt^2 = -g \quad (g > 0) \tag{5.138}$$

(with g as gravitational acceleration), while Newton's law is given by

$$d^2x/dt^2 = -g(1 + x/R)^{-2} \quad (R > 0) \tag{5.139}$$

in which, in addition to the (constant) g, the radius R of the earth (also constant) appears, and the centre of gravity is assumed to be at $x = -R$. Of course, this law has its origin in Newton's mechanics and theory of gravitation, and the explanation of (5.138) by these theories includes the explanation of g. But that is a matter of other, earlier reductions. Equation (5.139) marks the situation before the *last step* needed to explain Galileo's law. And that is what matters now. The core of the approximation here—corresponding to (5.135)–(5.136)—will be the additional assumption of small heights of fall

$$0 < x \ll R \, . \tag{5.140}$$

And, again, we see that this assumption does not restrict the set of solutions of (5.139), but the solutions themselves, namely, to their course in the (positive) neighbourhood of $x = 0$. Only the partial sections resulting from such a restriction are likely to be approximated by corresponding partial sections of Galileo's solutions (5.138). Globally, the solutions of (5.138) and (5.139) differ significantly. Among the latter, there are even cases where the body overcomes the gravitational acceleration and disappears into infinity, which is impossible with Galileo. But locally, above the

Earth's surface, even these solutions can be approximated in a Galilean way, just as all others.

Of course, this approximation does not take place in any interval on the line of fall, but only where the acceleration remains roughly the same according to Eqs. (5.138)–(5.139), namely, for (5.140). Equation (5.140) restricts the motion along with the height of fall to a time interval within which (5.140) indeed holds. This time, our reduction statement is, therefore:

– For every solution $x'(>0)$ of (5.139) there exists a solution x of (5.138), such that for every $\epsilon > 0$ there exists an $\delta > 0$, such that for all t, with

$$x'(t)/R < \delta \,,$$

– the following inequality holds.

$$|x'(t) - x(t)|/R < \epsilon$$

This statement obviously has the same structure as the corresponding statement for the radiation laws, and we will at once make a generalisation based on it.

Before that, however, we want to carry on the explanation of Galileo's law of falling another step. (For the further steps, we refer to Sect. 4.5.) The penultimate step is the explanation of the gravitational acceleration g in (5.139) and, thus, in (5.138). It comes about through the equivalence

$$\begin{aligned} N(x; m, R) \wedge g = m/R^2 &\vdash N_1(x; g, R) \\ N_1(x; g, R) \wedge m = gR^2 &\vdash N(x; m, R)\,, \end{aligned} \tag{5.141}$$

in which we put

$$N(x; m, R) \equiv d^2x/dt^2 = -m(R + x)^{-2}$$

and $N_1(x; g, R)$ stands for (5.139). The significant step is the first line. We learn to understand the gravitational acceleration by reducing it to the mass m and the radius R of the Earth in the sense of Newton's law of gravitation. We would be less inclined to say the same for the second line by interchanging m and g. Here, we have again the case, already discussed in Sect. 4.2, that the symmetry of a formal equivalence does not also apply to its content. In the present case, the symmetry in (5.141) could still be maintained insofar as the additional elements in both cases are simply empirical relations. The asymmetry only begins with the view that the mass m 'causes' the acceleration g, but not vice versa.

Local reductions can *generally* be described as follows. Under the pre-approximate assumptions made for asymptotic reductions (5.21)–(5.22), we additionally assume that extensions L' and L in M° are distinguished for the theoretical domains M' and M, respectively:

$$M' \subseteq L'; \quad M \subseteq L; \quad L', L \subseteq M^\circ \,. \tag{5.142}$$

As far as they do not belong to M' and M, respectively, the elements of L' and L are *local solutions* of the respective laws α' and α in (5.21), i.e., every $s' \in L'$ is subset of an $s_1' \in M'$ *and* solution of α', and accordingly in the unprimed case. Here,

$$L \cap L' = \emptyset \tag{5.143}$$

still holds. For Planck's law, e.g., the solutions restricted to finite intervals $(0, \nu)$ were added as such solutions with respect to (5.135). The actual vehicle of a local reduction is now a family of mappings

$$\Gamma_\delta : M^\circ \to M^\circ \text{ with } \Gamma_\delta(M') \subseteq L', \quad \Gamma_\delta(M) \subseteq L \tag{5.144}$$

which, depending on a real parameter $\delta > 0$, assigns to each solution from M' or M its δ-restriction in L' or L, respectively. Usually, here holds

$$\Gamma_\delta(s) \subseteq s\ . \tag{5.145}$$

For (5.135), these were the mappings assigning to each pair $\langle \rho, T\rangle \in M^\circ$ the pair $\langle \rho_\delta, T\rangle$, where ρ_δ is the restriction of ρ to the (T-dependent!) ν-interval

$$0 < \nu < (kT/h)\delta$$

and correspondingly for (5.133). Of course,

$$\Gamma_\delta(s) \subset \Gamma_\epsilon(s) \text{ for } \delta < \epsilon \tag{5.146}$$

has to hold, and the corresponding relation in the primed case. The approximation statement is then, in general,

$$\begin{aligned}&\text{For each } s \in M' \text{ there exists an } s \in M \text{ such that for each } u \in U^\circ\\ &\text{there exists a } \delta > 0 \text{ such that } \langle \Gamma_\delta(s'), \Gamma_\delta(s)\rangle \in u\ .\end{aligned} \tag{5.147}$$

Thus, the comparison does not take place by restricting M', as in the asymptotic reduction, but by restricting each solution of M' and of M.

This fact must be taken into account by briefly addressing now the question of *empirical progress* in this case as well. Indeed, e.g., the conservation of compatibility according to C_3 from Sect. 3.3 is very similar to the corresponding case of the asymptotic reduction in Sect. 5.1. Let us assume that the term t *itself*, but only the *elements* of the term s are measured data. Then the compatibility of data $a, \ldots, t^\circ$ with M reads:

$$\exists st. a \in u(s) \wedge \cdots \wedge \langle tt^\circ\rangle \in v \wedge \langle st\rangle \in M\ . \tag{5.148}$$

However, this condition would again be too weak, since in theory it allows for setting the measurement accuracy so high that our data are not also compatible with M'. We know this situation already from the asymptotic reduction, but to overcome it this time we have to proceed in a different way. We have to presuppose in (5.148) that we are dealing with restricted solutions

$$\exists st.a \in u(\Gamma_\delta^1(st)) \wedge \cdots \langle \Gamma_\delta^2(st), t^\circ \rangle \wedge \langle st \rangle \in M.\,, \tag{5.149}$$

but not with a restriction of the solution set M. Then we can fix the δ still left open here in the sense of *inverting* (5.147) with regard to M and M', i.e., in such a way that with given neighbourhoods u and v the equation

$$\begin{aligned} \langle st \rangle \in M \vdash \exists s't'.\langle s't' \rangle \in M' \wedge \langle \Gamma_\delta^1(s't'), \Gamma_\delta^1(st) \rangle \in U \\ \wedge \langle \Gamma_\delta^2(s't'), \Gamma_\delta^2(st) \rangle \in v. \end{aligned} \tag{5.150}$$

holds with

$$\langle f'f \rangle \in U \equiv f' \subseteq u(f) \wedge f \subseteq u(f')\,, \tag{5.151}$$

i.e., again with the uniform power. Here, too, as already with the asymptotic reduction, there seems to be no way of avoiding this additional assumption of also inverting (5.147), if we want to comply with the conservation laws (cf. the discussion in Sect. 5.1). With (5.148), from (5.149), follows

$$\begin{aligned} &\exists st.a \in u(f) \wedge \cdots \wedge \langle g, t^\circ \rangle \in v \\ &\quad \wedge \exists s't'.\langle s't' \rangle \in M' \wedge \langle f'f \rangle \in U \wedge \langle g'g \rangle \in v..\,, \end{aligned}$$

with obvious abbreviations by terms f, g, f', g', of which the dependence on s, t, s', t' has to be taken into account. We obtain from this by operating on the quantifiers:

$$\begin{aligned} &\exists s't'.\langle s't' \rangle \in M' \wedge \exists st.a \in u(f) \wedge \langle f'f \rangle \in U. \\ &\quad \wedge \ldots \\ &\quad \wedge \exists st.\langle gg' \rangle \in v \wedge \langle gt^\circ \rangle \in v.. \end{aligned}$$

The $\exists st$-terms containing u and U imply $a \in u^2(f')$ because of (5.151), and the $\exists st$-term containing v is essentially $\langle g't^\circ \rangle \in v^2$). So we get

$$\exists s't'.\langle s't' \rangle \in M' \wedge a \in u^2(\Gamma_\delta^1(s't')) \wedge \ldots \wedge \langle \Gamma_\delta^2(s't') \rangle \in v^2.\,, \tag{5.152}$$

and, if (5.145) holds, even

$$\exists s't'.\langle s't' \rangle \in M' \wedge a \in u^2(s') \wedge \ldots \wedge \langle t't^\circ \rangle \in v^2. \tag{5.153}$$

and, thus, compatibility with M' in the sense of (5.148) and (5.149), respectively.

It should not be concealed that here, as in all other cases involving measured data of different levels, we have used the uniform structure of the power according to (5.151) for general considerations, while in concrete examples involving spaces of functions we have provided uniform structures adapted to them. Of course, it must be checked whether these uniform structures amount to the same, at least topologically, or whether the conservation laws for the concrete cases also apply with more specific uniform structures. As in so many other respects, our explanations have to remain incomplete in this one as well.

References

Becker, R. [2]1966. *Theorie der Wärme*. Berlin: Springer. Engl. transl.: *Theory of Heat*. 2nd ed. revised by G. Leibfried. Berlin: Springer 1967.

Berry, M. 1994. Asymptotics, Singularities and the Reduction of Theories. In: D. Prawitz et al. (Eds.), *Logic, Methodology and Philosophy of Science*. IX. Amsterdam: Elsevier. 597–607

Bourbaki, N. [4]1965. *Éléments de Mathématique, Livre III: Topologie Générale*, Ch. 1 et 2, Paris: Hermann.

Cohen-Tannoudji, C., Diu, B. and F. Laloe. 1977. *Quantum Mechanics*, vol. 1. New York: Wiley.

Ehlers, J. 1986. On Limit Relations Between, and Approximate Explanations of Physical Theories. In: R. B. Marcus et al. (eds.), *Logic, Methodology and Philosophy of Science*, vol.7. Amsterdam: North-Holland, 387–404.

Lin, C.P., and L.A. Segel. 1994. *Mathematics Applied to Deterministic Problems in the Natural Sciences*. New York: Macmillan.

Møller, C. [2]1972. *The Theory of Relativity*. Oxford: Clarendon Press.

Querenburg, B. v. 1973. *Mengentheoretische Topologie*. Berlin: Springer.

Scheibe, E. 1973b. Die Erklärung der Keplerschen Gesetze durch Newtons Gravitationsgesetz. In: E. Scheibe and G. Süßmann (Eds.), *Einheit und Vielheit. Festschrift für G. F. v. Weizsäcker zum 60. Geburtstag*. Göttingen: Vandenhoeck & Ruprecht, 98–118. Engl. transl.: The Explanation of Kepler's Laws, in: Scheibe 2001, 306–323.

Schpolski, E.W. 1954. *Atomphysik*, Vol. 1. Berlin: Deutscher Verlag der Wissenschaften.

Vollmer, G. 1989. Darf man Falsches lehren? *Naturwissenschaften* 76, 185–193.

Chapter 6
Partial Reductions

In the introduction to his book 'Classical Charged Particles', Fritz Rohrlich noted that the foundations of a theory are often subject to drastic modifications, in contrast to its contingent predictions, which always remain valid within the achieved measurement accuracy. In the development of physics, therefore, a hierarchy of "lower-level theories" and their subsequent "higher-level covering theories" emerged. But, as a consequence of these necessary modifications, the requirement "that the lower-level theory be derivable from the covering theory" has to be restricted. This holds "not so much with respect to the axiomatic framework, which is, in general, not a special case of the framework of the covering theory, but with respect to certain basic equations and postulates which contain all the predictive power of the lower-level theory" (Rohrlich 1965, 5). Applied to the reduction theory advocated here, Rohrlich's remark is a plea for *partial reductions*: It is not necessary and often not even possible to reduce a well-established, but finally overcome physical theory *in toto* to its successor theory. Rather, the reduction only needs to reproduce the empirically proven part of that older theory, and this need not include all axioms, possibly not even a single axiom. In fact, many physics textbooks practise Rohrlich's view without explicitly expressing it. Thus, for example, it is said that classical mechanics turns out to be a limiting case of quantum mechanics under certain conditions. The direct reason for such talk, however, is usually only that this or that specific classical law turns out to approximate the corresponding specific results of quantum mechanics if certain parameters, such as temperature, frequency, energy, etc., take extremal values. This way of talking about *the* classical mechanics as a limiting case of *the* quantum mechanics would then be just a façon de parler to summarise many individual cases of 'small' reductions in a *merely aggregative* way into a general state of affairs. And it could be that concerning reduction we can achieve no more than a state of affairs arising in this way for certain cases. Moreover, a commitment to reduction would be needless if, so to speak, the lowest level of a theory to be reduced gets no physical interpretation at all.

E. Scheibe, *The Reduction of Physical Theories*, Fundamental Theories of Physics 207,
https://doi.org/10.1007/978-3-662-65000-4_6

In the statements quoted above, Rohrlich avoids taking sides with the consistent physical interpretation of a theory, as well as with its opposite. In other words, he neither recommends a realistic interpretation of the empirically inaccessible foundations of a theory, nor the rejection of such a step and, thus, the instrumentalist view of physical theories. And most physicists take a similar stance today. Some justify this neutrality by simply pointing out that we often do not have sufficient reasons for a decision. On the other hand, again and again, physicists are in search of such decisions. The best known and oldest case is the question of whether the structure of matter is discrete or continuous. Currently, we consider this question decided in atomistic terms. The last proponent of atomism who still had to argue on philosophical rather than empirical grounds was Boltzmann. Should we nowadays try to reduce full continuum mechanics to quantum mechanics? Even after deciding this matter, an instrumentalist point of view without any commitments remains possible. For instance, in an influential 1965 handbook article on continuum mechanics, we read that, contrary to a widely held view, "continuum physics presumes *nothing* regarding the structure of matter. It confines itself to relations among gross phenomena, neglecting the structure of the material on a smaller scale" (Truesdell and Noll 1965, 5–6). This is obviously a plea for a continuum mechanics that takes the mathematical treatment of matter as a continuum, and with it the logical basis of the whole theory, no longer seriously from a physical point of view. This is the consequence, here, of the knowledge acquired in the meantime, rather than abandoning the theory in favour of a quantum mechanics that reproduces the empirical successes of continuum mechanics. However, the authors' disappointment that the time has yet to come when quantum physicists will "turn back to gross phenomena and demonstrate that their theories do in fact predict them" (ibid.; cf. Chap. 4 of Vol. II[1]) seems to contribute to this conservative strategy. Either way, quantum mechanics will not be able to reproduce full continuum mechanics.

Apart from such ontological qualms, from a historical point of view, it would be astonishing if every physical theory ever established and corroborated were to be improved in a way that left it untouched *as a whole* when it is finally overcome. The progress of physics would then be purely cumulative, and we have already reported in Chap. 1 that neither the philosophers of science nor the physicists themselves believe in such a trouble-free kind of progress. In the meantime, we have also seen from the approximate reductions of the previous chapter that physical theories are more or less *corrected* by their successors. In particular, the local reductions show an obvious aspect of reproducing the reduced theory only partially. But this is not yet the aspect we now want to focus on. In a local reduction, the theory statement to be reduced still enters as a whole, and, due to the approximate character of the reduction, one could not even say *which* well-determined part of the theory to be reduced is recovered. For the partial reductions now to be addressed, this can very well be said. The fact that a

[1] *Classical Mechanics and Quantum Mechanics* (Chap. X of Scheibe 1999).

theory T' partially reduces a theory T means that it does not (totally) reduce T itself, in the usual sense, but one or another theory $T_1, T_2, \ldots$ that can be obtained from T by 'deduction'. Apart from the question of what "one or the other" is supposed to mean here, we do not want to be too specific about how the theories $T_1, T_2, \ldots$ may be obtained from T. At least in the examples, however, these conditions will be determined with sufficient precision to justify the expression 'partial' in contrast to 'total' reduction. The procedures leading from T to the $T_1, T_2, \ldots$ may, in particular, be (total) 'derivations'. For these, one will generally demand that they are exact, so that their conclusions still appear as genuine results of the initial theory. In principle, however, we could also think here of relaxing the approach to approximations.

In any case, partial reductions, in contrast to all other kinds of reductions dealt with here, are *no* longer procedures for which we claim the status of genuine reductions. This is also related to the fact that the question of empirical progress, which is so fundamental for us, does not require a separate treatment for partial reductions. For, in a certain sense, partial reductions appear to be traced back to total ones, and it is these total and always approximate reductions that make empirical progress possible in the sense discussed in the previous chapter.

6.1 The Closed Reduction Square

The best systematic introduction to the field of partial reductions is to treat cases of partial reduction in which a total reduction exists, such that recourse to partial reductions would not actually be necessary. However, since partial reductions also exist within the framework of total reductions, the latter can be controlled by the former, so to speak—in fact, even explained. The kind of explanation we are interested in is the *reduction square*. Suppose that Σ and Σ' are the axioms of a theory T and T', respectively, and that Σ is approximately (and totally) reducible to Σ'. If now a statement β can be deduced from Σ and perhaps additional assumptions, one will suppose that there is a statement β' obtainable in the same sense from Σ', which in the light of the (total) reduction of Sigma to Σ' is *the* improvement of β in Σ'. Insofar as β and β' can also be considered axioms of independent theories, β will now also be approximately reducible to β'. And, insofar as this is true, Σ will be *also partially* reducible, namely, just by the reduction of β to β'. We would, therefore, have the square (6.1) below with approximate reductions in the horizontal direction and strict deductions, possibly even exact 'forward' reductions, in the vertical. This square would be closed in that both paths from β to Σ'—the one via β' and the other via Σ—would be viable and, so to speak, 'fit' each other. If (exact) reductions also take place in the vertical directions, with (6.1) we would have further cases before us in which a theory (here: β) would be reduced in two ways to another theory (here: Σ'), cf. Sects 4.5 and 5.2. Even more, the reduction from β to β' would be better understood by the deeper or more far-reaching one from Σ to Σ': the latter explains the former. Finally, the closed reduction square prepares us for the open one, in which the total reduction from Σ to Σ' does not exist or is not known and we only have the reduction in the lower row between the 'conclusions' β and β' at our disposal.

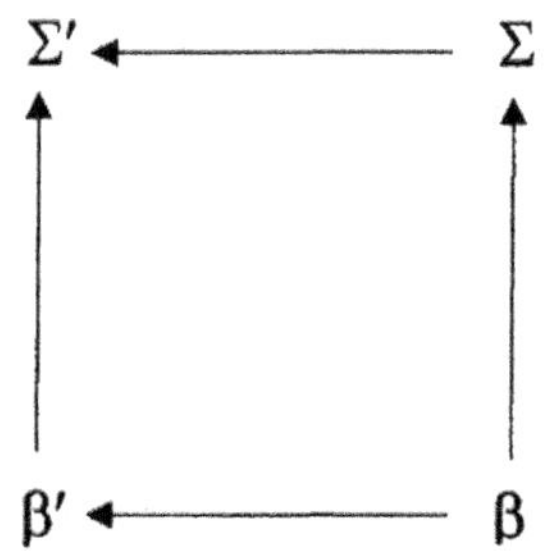

(6.1)

For a first example, let us return to the gas laws from Sect. 5.1, where we asymptotically reduced the ideal gas equation (Σ) to the van der Waals equation (Σ'). The ideal gas equation is now, in turn, a 2nd kind generalisation of Boyle and Mariotte's law (β)

$$pv = RT^\circ \tag{6.2}$$

with *fixed* temperature T°. The same applies to the van der Waals equation with respect to the equation (β')

$$(p + a/v^2)(v - b) = RT^\circ \,. \tag{6.3}$$

It is now clear that the claim of the asymptotic reduction of (6.2) to (6.3) can be obtained directly from the corresponding claim (5.13) of Sect. 5.1 with variable T. It is also obvious that (6.3) plays the role within the van der Waals theory that (6.2) has in relation to the ideal gas equation. Finally, it is clear that the specialisation of the van der Waals equation to (6.3) corresponds exactly to the specialisation of the ideal gas equation to (6.2). So we face a closed reduction square under ideal conditions. The latter are explicitly mentioned here despite their triviality, because it is precisely these that become more and more questionable in more complicated cases.

The following example is also simple. Here, we will start from the lower reduction level and only then ascend to the upper one. Among the classical cases of asymptotic reduction in Sect. 5.1, we could also have included the reduction of Kepler's 3rd law (β)

$$a^3/T^2 = (1/4\pi^2)m_1 \tag{6.4}$$

to its Newtonian improvement (β')

$$a^3/T^2 = (1/4\pi^2)m_1(1 + m_2^\circ/m_1)^{-2} \,. \tag{6.5}$$

With fixed m_2° and variable m_1, a, and T, (6.5) turns asymptotically into (6.4) under the constraint

$$m_2^\circ/m_1 \ll 1 \,. \tag{6.6}$$

Of course, such an explanation of (6.4) by (6.5) via (6.6) will remain unsatisfactory as long as we do not know how to arrive at (6.5) instead of (6.4). The correction term in (6.5) compared to (6.4) could look different in many ways without impeding the asymptotic reduction. In the present case, we do not only know the origin of (6.4) and (6.5) from Kepler's and Newton's theory (Σ and Σ' respectively), but we also dispose of an asymptotic reduction of the former to the latter (cf. (5.32)–(5.40) of Sect. 5.1). Thus, we can control the reduction of (6.4) to (6.5) in all details. To exclude further reductions, we assume for this procedure that Kepler's and Newton's theories are restricted by the further conditions that we operate in the centre of gravity system and consider only periodic solutions, in addition to (5.32) and (5.33) and (5.40).

In (4.82)–(4.86) of Sect. 4.4, we got to know Kepler's third law as a coarsening of Kepler's theory, the latter therefore as a refinement of the former. Correspondingly, Newton's improvement (6.5) turns out to be a coarsening of Newton's theory of gravitation for two masses. This is precisely what the correction term explains. In this case, we thus obtain the connection from the lower to the upper level in (6.1) both on the left and on the right by the exact reduction we called "refinement". We can then apply the 'control' mentioned above, as follows. Given an accuracy $\epsilon > 0$ for the approximation of a solution $\langle m_1, a, T\rangle$ of (6.5) under the constraint (6.6) by a solution $\langle m_1', a', T'\rangle$ of (6.4), we first proceed from $\langle m_1, a, T\rangle$ to a solution $\langle m_1, \mathbf{x}_1, \mathbf{x}_2\rangle$ of Newton's theory (cf. (5.32) and (5.33) and (5.40) of Sect. 5.1), which yields the given values of a and T. Since we are dealing with a refinement, such a solution always exists. This $\langle m_1, \mathbf{x}_1, \mathbf{x}_2\rangle$ is then approximated by a Kepler solution $\langle m_1', \mathbf{x}_1', \mathbf{x'}_2\rangle$ according to (5.36) and (5.37). It turns out that the former requirement for the choice of $\mathbf{x}_1'$ and $\mathbf{x}_2'$ leads to values $a' = a$ and $T' = T$, respectively, since $\mathbf{x}_1' = \mathbf{x}_2'$ should hold. On the other hand, the choice of m_1' just corresponds to the correction term in (6.5).

In order to make clear what makes the indicated control possible, it is useful to start from the question of what guarantees that one obtains β from Σ in the *same* way as β' from Σ'. After all, even if Σ and Σ' are related by reduction, we can initially obtain β and β' from Σ and Σ', respectively, in a completely uncorrelated way. For example, β could be Kepler's third law, but β' (in Σ') the first, i.e., the statement that the second body is moving on an ellipse. Then, it would make no sense to expect a reduction of β to β'. The question is under which conditions for β and β' we could justify expecting it. We could if β with respect to Σ played the same role as β' with respect to Σ'. What this means can be specified as follows for the case that in (6.1) we have an asymptotic reduction at the top and a refinement on the left and on the right. Starting point is the relation of Σ and Σ' established at the end of the introduction to Chap. 5, whereby according to Sect. 5.1 we have to add that the same variables occur in both theories, i.e., the s_1 are omitted. It is easy to see that, under these conditions, it makes sense to speak of the two refinements being the same, i.e., having the same mapping terms q and q_1 (cf. (4.92–4.95) of Sect. 4.4).

As a third example of the closed reduction square, let us take up the plane pendulum discussed in Sect. 5.3, in which the exact equation of motion

$$d^2\varphi/dt^2 + \omega^2 \sin\varphi = 0 \ \ (\text{for } \Sigma') \tag{6.7}$$

asymptotically reduces the oscillator equation

$$d^2\varphi/dt^2 + \omega^2\varphi = 0 \ \ (\text{for } \Sigma) \tag{6.8}$$

under the condition of small amplitudes $\alpha \ll \pi$. This relation can be used to explain a reduction relation between two frequency formulae. Namely, (6.7) and (6.8) are refinements of

$$\tau = (1/\omega)\, 4 \int_0^{\pi/2} (1 - \sin^2(\tfrac{1}{2}\alpha)\sin^2\gamma)^{-1/2} d\gamma \tag{6.9}$$

and

$$\tau = (1/\omega)\, 2\pi\ , \tag{6.10}$$

respectively, with τ as the oscillation period and α as the amplitude of φ. For (6.7) and (6.9), we had seen this in Sect. 4.4, and accordingly it holds that (6.8) is a refinement of (6.10), τ in (6.10) being now independent of α. On the other hand, (6.10) turns out to be asymptotically reducible to (6.8) in such a way that, again, a closed reduction square is obtained: $\langle\tau, \alpha\rangle$ with (6.10) is explained by a solution $\langle\tau', \alpha'\rangle$ of (6.9) with given precision, in that there exists a solution φ of (6.8) and a solution φ' of (6.7) such that (i) they induce those values, and (ii) φ' explains φ with corresponding precision.

Returning to the Kepler-Newton relation, we notice that, strictly speaking, we have not used the reduction developed in (5.32)–(5.40) in the upper line of the reduction square just established to demonstrate that relation. Rather, this time, we have assumed that we are operating in the centre of gravity system and that only periodic solutions are permitted. In fact, Kepler's laws and, especially, the third law are only meaningful under these constraints. In contrast, (5.32)–(5.40) of Sect. 5.1 dealt with the Galilei-invariant versions of the two theories, admitting also non-periodic motions. It is now easy to see that we can sort this incorrectness out by another reduction square sitting on top of the first square:

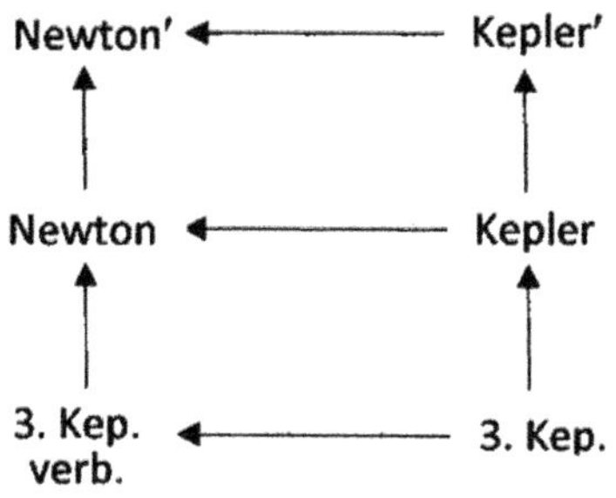

(6.11)

In this double square, all horizontal arrows stand for asymptotic reductions, the lower vertical ones for the refinements already examined, and the newly added upper ones for direct generalisations. For these, we must now ensure that they are 'the same' on the left and on the right, and this time this can no longer be achieved literally, but only analogously. On the one hand, we want to go into the frame of reference in which the orbit of the planet appears as a conic section. In Newton's theory, this would be achieved by assuming that the centre of gravity of the system is at rest. In the conceptual world of Kepler's theory, however, there is no centre of gravity. In this case, the sun would have to be at rest. If we wanted to continue to describe the elliptical orbit of the planet, we would again have difficulties if we wanted to achieve this through a condition of energy. For the concept of energy is also foreign to Keplerian theory. Here, we can help ourselves by requiring in both cases that we are dealing with a periodic motion. All in all, however, equality in the literal sense can no longer be established here, and this difficulty will continue to accompany us without enabling us to resolve it.

6.2 The Open Reduction Square

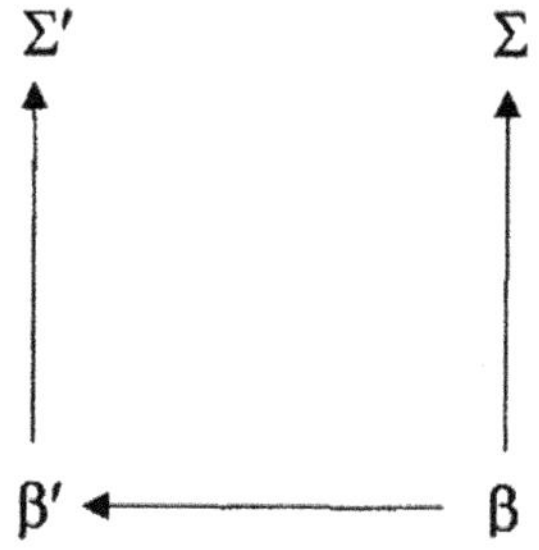

(6.12)

The considerations of the previous paragraph served only to prepare the study of 'real' partial reductions. We understand these to be reduction squares open at the top (6.12). Here, the specification of a direct reduction of Σ to Σ' is missing. We only learn that a 'result' β of Σ is (approximately) reducible to a 'result' β' of

Σ'—however, here, too, such that β' is *that* 'result' of Σ' corresponding to β in Σ and thus, likewise, is *the* improvement of β in Σ'. In this sense, Σ would only be *partially* reduced to Σ', namely, by reducing β to β' in their roles within Σ and Σ', respectively. Omitting the direct reduction of Σ to Σ' may be due to the fact that none is known, or even that none exists. A more profound theory of partial reductions would actually have to start with the second case, and then of course by *proving* the impossibility of reduction. This would be the strongest motive for the formation of such a theory. But it is obvious that proofs of this kind are expected to be difficult to impossible. So, with regard to the examples, we will (and must) content ourselves with considering cases in which the closure of the square (6.12) is uncertain or would even be a daunting task.

One such case is the reduction of classical mechanics (CM) to quantum mechanics (QM). It is obvious that one can decompose this reduction into one from CM to classical statistical mechanics (CSM) and a subsequent one from CSM to QM. We will deal with these problems in more detail in Chap. 4 of Vol. II. For the moment, the question is whether there are not at least partial reductions from CSM to QM. Concerning it, we already been said in the introduction to this chapter that most answers encountered in the literature are, if not verbatim, then in substance no more than the mere claim that partial reductions exist. The first example we want to look at more closely is a partial reduction within the classical and quantum mechanical theory of the harmonic oscillator. In Sect. 6.1, we had already compared the results of the classical and quantum mechanical treatment of an ensemble of oscillators in thermal equilibrium. The classical mean energy and the spatial variance, as well as the spatial distribution itself, could be asymptotically reduced to the corresponding quantum mechanical values. We now deal with the additional claim that this, in fact, concerns partial reductions of CSM to QM (cf. Cohen-Tannoudji et al. 1977, 620–623).

This claim amounts to saying that, for example, the quantum mechanical formula for the mean energy is obtained from QM in either *the same* or a completely *corresponding* way as the corresponding classical formula is obtained from CSM. Since, in the present case, the reversal of this derivation is an exact reduction, one may also say: We are dealing with the question of what it means that the two formulae can be reduced to the general theories either in the same way or in a corresponding manner. We have already emphasised in the previous section that this question has to be asked and that finding an answer may be difficult. These difficulties mainly arise because we miss control of the equality in question by total reduction which, in turn, is due to the structural differences between CSM and QM. Nevertheless, it is claimed that, e.g., with (5.15) of Sect. 5.1 one has found a better formula for the same state of affairs as with (5.14) and, accordingly, for (5.17), (5.18), (5.19) and (5.20) there. How can this be justified?

Of course, this can only be done with a precise analysis of the two exact reductions in question. These are also interesting to us because they show an accumulation of iterations that illustrates well the main thesis of this study, that reductions are obtained synthetically. Although only two different kinds of reductions are overall involved—a refinement and a subsequent improper generalisation—, the refinement is, again, decomposed into three other refinements, including an extension, and the improper generalisation is decomposed into two reductions of the same kind. At first, however, the quantum-mechanical justification of the classical formula for the mean energy (as well as that for the spatial variance) allows a further asymptotic reduction to the corresponding quantum-mechanical result. In the end, we arrive here at a total of six consecutive reductions of three different kinds. Those who still consider this procedure to be a reduction at all will easily come to realise how inappropriate and difficult a conception of reduction would be that does not take the path of combining reductions. We will now run through the two reduction chains in parallel for the case of the spatial variances. For better orientation, we distinguish six levels, one after the other, which indicate the respective stage of the reduction.

At level 1, we have the two formulae (5.17) and (5.18) of Sect. 5.1 for the relation of the spatial variance Σ to the mass m, the frequency ω of the oscillator, and the temperature T. By *refinement*, we get to level 2 to the formulas (5.19) and (5.20) for the spatial distributions. Since we also know an asymptotic reduction for them, we still face a closed reduction square up to this point. The likewise asymptotic relationship between the variances can be understood by the corresponding relation between the spatial distributions themselves. And there can also be no doubt that the two refinements are equal. They are provided by one and the same relation

$$\sigma^2 = \int q^2 \, \mathrm{pr}(q)dq - \left\{ \int q \, \mathrm{pr}(q)dq \right\}^2 \tag{6.13}$$

between the variance and the respective probability distribution. The difficulties begin with the next step, that is, the *refinement* of the spatial distributions to the complete state densities. Already here, the hitherto common foundation comes apart at the seams. For the classical case, at level 3, we obtain the phase space of the oscillator and, on it, the density of states ρ with

$$\rho(q, p) = \exp\{(-1/kT)(1/2m)p^2\} \exp\{-(1/kT)(m/2)\omega^2 q^2\} \,. \tag{6.14}$$

In the quantum mechanical case, we get the Hilbert space of the oscillator with the operators Q and P for position and momentum, and the density of states (statistical operator)

$$\rho_{\mathrm{qu}} = \exp\{(-1/kT)(1/2m)P^2\} \exp\{-(1/kT)(m/2)\omega^2 Q^2\} \,. \tag{6.15}$$

The mapping term leading from (6.14) and (6.15) to (5.19) and (5.20) of Sect. 5.1, which transforms ρ into pr, or ρ_{qu} into pr_{qu}, can still be obtained by a common superior definition, which, however, in concreto is already to be applied to quite different structures. The superior definition is (cf. Mackey 1963, Chap. 22) that a real probability distribution pr_ρ^A is assigned to each state ρ and each magnitude A, which indicates for a value set a of A how probable it is that A in state ρ has a value in a. If A, e.g., is specified as a position Q, as is the case here, the resulting mapping assigns a spatial distribution to each state ρ. However, the concrete rules in phase space or Hilbert space are different:

$$pr(q) \text{ resp. } pr_{\text{qu}}(q) = \begin{cases} \int \rho(q, p)dp & \text{(CSM)} \\ Tr(\rho_{\text{qu}} F_q^Q) & \text{(QM)}, \end{cases} \tag{6.16}$$

where F_q^Q is the spectral decomposition of Q and Tr is the operator trace. From (6.14) or (6.15) and (6.16) follow (5.19) and (5.20) of Sect. 5.1. But now, for the first time, we no longer have a closed square. In order to get it, we would have to be able to (approximately) reduce (6.14) to (6.15). And that has to remain an open matter for the time being.

The next step now brings us by an *extension* to level 4 where, in addition, a Hamiltonian function H or operator H_{qu} is introduced, so that *instead of (6.14)* and (6.15)

$$H(q, p) = (1/2m)p^2 + (m/2)\omega^2 q^2; \quad \rho(q, p) = \exp\{-(1/kT)H(q, p)\} \tag{6.17}$$

holds classically and

$$H_{\text{qu}} = (1/2m)P^2 + (m/2)\omega^2 Q^2; \quad \rho_{\text{qu}} = \exp\{-(1/kT)H_{\text{qu}}\} \tag{6.18}$$

holds quantum mechanically for it. It is obvious that (6.14) and (6.15) follow from (6.17) and (6.18) and that we are dealing with a conservative extension according to Sect. 4.5. Again, we cannot reduce (6.17) to (6.18), so that our "reduction tower" remains open at the top. Moreover, because of the difference of the cores, i.e., essentially the phase space and Hilbert space, the two extensions are only comparable by analogy. The formal analogy between (6.17) and (6.18) is obvious. But it is a purely mathematical one, which is not supported by any correspondence in content. The fact that we are still talking about the same thing in both theories is mainly due to the similarity of the formulae. This situation is, of course, also caused by the strongly mathematical character of the assumptions in (6.17) and (6.18); cf. what was said about it in Sect. 2.3.

The two assumptions in (6.17) and (6.18) are very special. To get rid of them, we now have to make two improper generalisations. We enter level 5 by omitting the assumptions (6.17) and (6.18) on the left, along with the oscillator parameters m and ω. This is an improper generalisation, in the sense of Sect. 4.5, to a theory in which the Hamiltonian function or operator is already quite general, but not yet the density of states. The latter is the density of states of a canonical totality of temperature T and, as such, stationary in time. It should be recalled here that Σ or σ_{qu} and pr or pr_{qu} should, in fact, be conceived as functions of time, yet in the special case of a canonical totality they are time-independent. So now also ρ or ρ_{qu}. Therefore, before the last reduction, we can add the equations of motion

$$d\rho/dt = \{H, \rho\} \quad \text{(Poisson)} \tag{6.19}$$

or

$$\hbar i(d\rho_{\text{qu}}/dt) = [H, \rho] \quad \text{(Commutator)} \tag{6.20}$$

to the assumptions (6.17) and (6.18) on the right, which the ρ or ρ_{qu} from (6.17) or (6.18) satisfies (cf. (6.17) and (6.18)) on the right); and (6.19) and (6.20) is equivalent to (6.17) and (6.18) on the right. An improper generalisation follows, in which we omit (6.17) and (6.18) on the right and, in addition, T. This brings us, then, to level 6 of the general CSM or QM.

We find partial reductions also *within* the large framework theories CM and QM, and they can be associated with considerable mathematical difficulties, if not so much conceptual ones. Such a case is the sometimes occurring double treatment of the theory of hydrogen-like atoms in QM (cf. Blochinzev 1953, Sects. 49 and 104). The first treatment (T) deals with the motion of an electron in the Coulomb field of the nucleus, which is assumed to be at rest. In the second treatment (T'), the motion of the nucleus is also taken into account. What is not done at all, however, is to clarify how these two theories relate to each other. And that would indeed be quite difficult. Instead of a total reduction, which may well be assumed here, we see a partial reduction of the energy formula for the case of the nucleus at rest to the corresponding formula for the case of the nucleus in motion. This is, of course, based on a previous decision as to how the energy is defined in both cases. And, here, we face the conceptual difficulty that the difference between the theories also forces us to use different definitions which, nevertheless, mean the same.

First of all, our problem also contains that of the relation between CM and QM, given that theory T, with its assumption of a nucleus at rest, makes a classical assumption that contradicts QM. In contrast, in T', the nucleus is treated by quantum mechanics. However, this difficulty hardly becomes manifest because in T the nucleus has disappeared, replaced by the Coulomb field. The comparison, thus, focuses on comparing a 1-particle theory (T) with a 2-particle theory (T'). The original problem of a total reduction of T to T', thus, looks as follows. In addition to the mass m_2 and charge e of the electron, the 1-particle theory T (the theory of the second particle, from the point of view of T') has, as parameters, the Hamiltonian (in spatial representation)

$$H_2 = -(\hbar^2/2m_2)\nabla_2^2 - Ze^2/r_2 \tag{6.21}$$

(Z the nuclear charge number and r_2 the 'distance of the electron' from the centre) and the density of states W_2 (statistical operator), which as a function of time satisfies the equation

$$i\hbar(dW_2/dt) = H_2W_2 - W_2H_2\,. \tag{6.22}$$

As a 2-particle theory, T' has, in addition to the masses m_1 and m_2, as well as charges Ze and e, the Hamiltonian

$$H = -(\hbar^2/2m_1)\nabla_1^2 - (\hbar^2/2m_2)\nabla_2^2 - Ze^2/r_{12} \tag{6.23}$$

(r_{12} the 'distance of the electron from the nucleus' that is now moving along) and the density of states W with the equation

$$i\hbar(dW/dt) = HW - WH \tag{6.24}$$

corresponding to (6.22). The connection between T and T' is now established by uniquely assigning to W the density of states Γ(W) through the condition

$$\mathrm{Tr}\{W(1 \otimes F_2)\} = \mathrm{Tr}(\Gamma(W)F_2)\ , \tag{6.25}$$

required for all F_2 (where F_2 is an arbitrary projection operator on the Hilbert space of the second particle). $W(t)$ stands for the knowledge at the moment about the 2-particle system. According to T', Γ(W) is the knowledge contained in W about the second particle—the electron. This Γ(W) could, therefore, be compared with W_2 of (6.22). Initially, we only know that what the dynamics Γ(W) has to satisfy is controlled via (6.25) by the dynamics of W according to (6.24). A (total) reduction would be achieved here if it were possible to reduce T approximately to the theory connected with Γ(W). This would then be followed by an exact reduction based on (6.25), which we introduced in Sect. 4.4 as a refinement, with some reservations.

We do not undertake this total reduction here. It is likely to be accompanied by several difficulties. However, we can easily arrive at a partial reduction as follows. The energy formula

$$E_n = -(Z^2e^4m_2/2\hbar^2)(1/n^2) \tag{6.26}$$

follows from the theory T. In contrast, the formula

$$E_n' = -(Z^2e^4m_R/2\hbar^2)(1/n^2) \tag{6.27}$$

is derived from T' (cf. (5.92) and (5.93) of Sect. 5.2). It follows from (6.26) by replacing m_2 by the so-called reduced mass

$$m_R = m_1m_2/(m_1 + m_2)\ . \tag{6.28}$$

Since m_r approaches m_2 for $m_1 \to \infty$, it is no problem to identify (6.26) as limiting case of (6.27), in the sense of Sect. 5.2, for an 'infinitely heavy' nucleus. So the lower level of the open reduction square (6.12) is established. However, some more consideration is needed to obtain also the vertical arrows of (6.12), i.e., to determine the exact reduction of (6.26) to (6.21), (6.22) and of (6.27) to (6.23) and (6.24), by making clear that (6.27) is the improvement of (6.26) in T' and, in particular, that (6.26) and (6.27) actually state the same thing.

Two points need to be briefly addressed here. One concerns the direct origin of the energy formulae (6.26) and (6.27). For (6.26), it is clear that the E_n are simply the eigenvalues of the Hamiltonian operator (6.21). Not so clear is the origin of the E_n' in (6.27). Unlike in T, which excludes a translational motion of the system (Coulomb field + electron) is excluded, T' again admits such a motion for the system (nucleus +

electron). Therefore, we have to subtract the translational part from the total energy (6.23) in order to arrive at the "inner" energy corresponding to (6.21). We do so by splitting into centre of gravity and relative coordinates, and the inner energy we are looking for then concerns only the latter. This splitting leads next to (6.28) to the total mass

$$m_S = m_1 + m_2 \tag{6.29}$$

and by converting (8a) to

$$H = H_S \otimes' \mathbf{1} + \mathbf{1} \otimes' H_R \tag{6.30}$$

with

$$H_S = -(\hbar^2/2m_S)\nabla_S^2, \quad H_R = -(\hbar^2/2m_R)\nabla_R^2 - Ze^2/r_R \tag{6.31}$$

($\otimes'$ in (6.30) denotes the product with respect to the centre of gravity and relative coordinates, not the particle coordinates). The E_n' are then the (discrete) spectrum of the internal energy H_R.

The other question now is how to get from the description of our system by W_2 or W to the description by n and E_n or n and E_n' (besides the masses and charges). For T, this way consists in the *specialisation* of W_2 to a stationary state

$$\psi_2 \exp\{-(i/\hbar)E_n t\} \text{ with } H_2\psi_2 = E_n\psi_2 \tag{6.32}$$

and subsequent coarsening to the mere value E_n for a certain n (since the energy in the Coulomb field is degenerate). For T', there is first also a specialisation of W to a pure product state $\varphi \otimes' \psi$, where ψ is a stationary state

$$\psi_R \exp\{-(i/\hbar)E_n' t\} \text{ with } H_R\psi_R = E_n'\psi_R\,. \tag{6.33}$$

A twofold coarsening follows, in which first φ is projected away and then n and E_n' are obtained as above. We thus see that here too, in the opposite direction, there are exact reductions of (6.26) to (6.21), (6.22) and (6.27) to (6.23) and (6.24), which analogously correspond to each other.

Investigations that do not only concern the energy formulae have been carried out for the simple model case of the 2-particle oscillator

$$H = -(\hbar^2/2m_1)\nabla_1^2 - (\hbar^2/2m_2)\nabla_2^2 + (k/2)(q_1 - q_2)^2 \tag{6.34}$$

in relation to the (usual) 1-particle oscillator

$$H_2 = -(\hbar^2/2m_2)\nabla_2^2 + (k/2)q_2^2 \tag{6.35}$$

(cf. Scheibe 1981a and the classical case in Sect. 5.3). The question was: What are the improvements of the formulae for the position and momentum densities of the stationary states of (6.35) in the theory (6.34)? To answer it, we first have to clarify from which states of the theory governed by (6.34) the improved position and momentum densities should be calculated. We first take the decomposition (6.28), (6.29), (6.30) and (6.31), which here leads to

$$H_S = -(\hbar^2/2m_S)\nabla_S^2; \quad H_R = -(\hbar^2/2m_R)\nabla_R^2 + (k/2)q_R^2 \tag{6.36}$$

in addition to (6.28), (6.29) and (6.30). With respect to this decomposition, we now choose a product state

$$\varphi = \varphi_S \otimes' \varphi_R \tag{6.37}$$

with

$$\hbar i(d\varphi_S/dt) = H_S\varphi_S; \quad \hbar i(d\varphi_R/dt) = H_R\varphi_R \tag{6.38}$$

under the following additional conditions: φ_R is stationary. Similar to the energies, the stationary states of H_R improve the corresponding states for H_2. In addition, however, there is the condition for φ_S that the centre of gravity on average is at rest in the point of origin (corresponding to the fixed nucleus) and that φ_S is a 'minimal packet', i.e., its Heisenberg uncertainty relations are equations. Then, φ_S is uniquely determined (as a Gaussian distribution) by its initial (spatial) variance Σ.

Under these presuppositions, the spatial densities

$$|\varphi_2(x_2)|^2 \tag{6.39}$$

(and similarly the momentum densities) of the stationary states of H_2 turn out to be limiting cases of the spatial densities

$$\int |\varphi_S(m_1 x_1 + m_2 x_2/m_s, t)|^2 |\varphi_R(x_2 - x_1, t)|^2 dx_1 \tag{6.40}$$

(and correspondingly of the momentum densities) of the states Φ under the given conditions. However, this time, the limiting condition of an infinitely heavy nucleus is not sufficient. To $m_1 \to \infty$, a limit must be added that expresses the classical behaviour of the centre of gravity (at rest), that is, $\sigma \to 0$ for the variance of φ_S. However, even this is not sufficient, insofar as the decay of the wave packet φ_S needs to be delayed as much as possible for the approximation. With the time-dependent mean variance of φ_S,

$$\sigma^2(t) = \sigma^2 + \hbar^2(t - t_0)^2/4m_S^2\sigma^2\,, \tag{6.41}$$

we finally see that

$$m_1 \to \infty, \quad \sigma \to 0, \quad m_1\sigma \to \infty \tag{6.42}$$

is altogether sufficient. Equation (6.41), however, also demonstrates that this is, at best, a 'pseudo-local' limiting case reduction (cf. Sect. 5.3). Whereas the spatial probabilities (6.39) are time-independent, the spatial probabilities (6.40) depend on time, and the time dependence of (6.41) makes it clear that our approximation only holds for finite time intervals. However, these can be arbitrarily long if only $m_1\sigma$ is sufficiently large.

Recently, these investigations have been carried out much further with considerable mathematical effort (cf. Osterfeld 1988). Let us mention only one typical result. It directly concerns our 'original problem' of how the states $W_{2,t}$ evolving according to (6.22), relate to the projections $\Gamma(\mathrm{W_t})$, with W_1 evolving according to (6.24), formulated above under (6.21)–(6.25). If we introduce (making the m_1-dependence explicit) the unitary operators

$$U_{2,t} = \exp\{iH_2t\}, \quad U_t(m_1) = \exp\{iH(m_1)t\} \tag{6.43}$$

which bring about this development, then it would be a matter of comparing

$$U_{2,t}\Gamma(W_0)U^*_{2,t}$$

with

$$\Gamma(U_t(m_1)W_0U_t(m_1)^*) ,$$

with the initial values W_0 and $\Gamma(\mathrm{W_0})$, cf. (6.25). The respective result may be expressed by means of the unitary operators:

$$(C_\lambda\psi)(x_1, x_2) = \lambda^{1/2}\psi(\lambda x_1, x_2) . \tag{6.44}$$

For

$$m_1 \to \infty, \quad \lambda \to 0, \quad \lambda^2 m_1 \to \infty , \tag{6.45}$$

$$\Gamma(U_t(m_1)C^*_\lambda W_0 C_\lambda U_t(m_1)^*) \to U_{2,t}\Gamma(W_0)U^*_{2,t} \tag{6.46}$$

holds in the trace norm (ibid., p. 42). We have to leave it open the whether this result means to close the open reduction square described above.

In this section, we have always assumed that obtaining β' from Σ' and β from Σ in the sense of (6.12) is an exact procedure. In this way, β' and β seemed to be genuine results of the theories Σ' and Σ, respectively, and this is worth attaching importance to in a notion of reduction that is already substantially weakened. At least in certain cases, however, one could also think of a further generalisation without too many qualms, considering also the relation between approximate results as contributions to a partial reduction of Σ to Σ'. For example, Einstein provided a formula for the specific heat of a crystal on a quantum theoretical basis, which in the limiting case of high temperatures turns into the classical formula. Einstein's result has since been improved by Debye and, further, within a genuine lattice theory (cf. Becker 1966, V. A). And, of course, these improvements also contain the classical limiting case. So why should all these theories, although obtained only approximately, not be regarded as contributing to the partial reduction of classical statistical mechanics to quantum mechanics? A closer examination of this question would, however, necessitate a much more detailed treatment of the other question already raised: How are we to guarantee, in general, that the 'derivations' of β' from Σ' and of β from Σ, from stage to stage, are still about the same thing, albeit within the framework of different, and mutually incompatible, theories? No such clarification seems possible at present and, therefore, we must also leave the other question open. For the same reason, we could not give a precise concept of partial reduction in general. For particularly difficult cases, however, we return to these questions in Volume II.

References

Becker, R. ^{2}1966. *Theorie der Wärme.* Berlin: Springer. Engl. transl.: *Theory of Heat.* 2nd ed. revised by G. Leibfried. Berlin: Springer 1967.

Blochinzev, D. I. 1953. *Grundlagen der Quantenmechanik.* Berlin: Deutscher Verlag d. Wissenschaften.

Cohen-Tannoudji, C., B. Diu, and F. Laloe 1977. *Quantum Mechanics*, vol. 1. New York: Wiley.

Mackey, G. W. 1963. *Mathematical Foundations of Quantum Mechanics.* New York: Benjamin.

Osterfeld, T. 1988. *Beispiel einer approximativen Theoriebeziehung innerhalb der Quantenmechanik.* Diploma thesis, University of Osnabrück.

Rohrlich, F. 1965, ^{3}2007. *Classical Charged Particles.* Addison & Wesley: Reading, MA. Quoted after the third edition, Singapore: World Scientific 2007.

Scheibe, E. 1981a. Eine Fallstudie zur Grenzfallbeziehung in der Quantenmechanik. In: J. Nitsch et al. (Eds.), *Grundlagenprobleme der modernen Physik. Festschrift für P. Mittelstaedt zum 50. Geburtstag.* Mannheim: Bibliographisches Institut 257–269. Engl. transl.: A Case Study Concerning the Limiting Case Relation in Quantum Mechanics. In: Scheibe 2001, 339–351.

Truesdell, C., and W. Noll. 1965. The Nonlinear Fieldtheories of Mechanics. In: S. Flügge (Ed.), *Handbuch der Physik* III/3. Berlin: Springer.

Appendix A
Table of Contents Volume II

This table of contents corresponds to the actual contents of Part II of the book published in 1999. The original plan given at the end of the German edition of the present book was more comprehensive (see also the corresponding footnotes in Chaps. 1 and 3):
7. Reductions to Relativistic Theories
8. Newton's and Einstein's Theories of Gravitation
9. Classical Mechanics and Quantum Mechanics
10. Micro Reductions
11. Reductions to Field Theories
Appendix: Relation to G. Ludwig's Theory of Reduction

E. Scheibe, *The Reduction of Physical Theories*, Fundamental Theories of Physics 207,
https://doi.org/10.1007/978-3-662-65000-4

Erhard Scheibe (1927–2010)

Appendix B
The Legacy of Erhard Scheibe

Brigitte Falkenburg

I.

Erhard Scheibe was born in Berlin on September 24, 1927, a few days after the physicist Niels Bohr gave the famous Como lecture which sparked the Bohr-Einstein debate on the nature of quantum mechanics.[1] This was something of an omen, for philosophical problems concerning the relations between quantum mechanics and classical physics stood at the centre of Scheibe's work throughout his life.

The family home did not guide him in this direction. Erhard was the only child of Albert Scheibe and Maria, née Heidenreich, and grew up in an upper middle-class ambience marked by discipline and subtle taste. His father, a former naval officer, became a business man after 1918, and contributed a preface to the 1937 edition of Admiral Scheer's book on the German armada in the First World War. Erhard's uncle on the father's side was the sculptor Richard Scheibe, a member of the Berlin Secession, who taught at Berlin University of the Arts from the mid 1930s. Gerhard Marcks, one of Richard Scheibe's friends, made a portrait sculpture of the father Albert.

The adolescent Scheibe belonged to that small group of German youth who were forced to adulthood early, and yet largely escaped the utter destruction brought about by the course of the Second World War. Erhard attended the Mommsen Gymnasium in Berlin-Charlottenburg until the age of fifteen, when he became a Flakhelfer, an anti-aircraft auxiliary. From the autumn of 1944 until the spring of 1945, he managed to avoid the military draft by continually changing his residence. He used to recount that in the summer after the end of the Second World War, he read Kant's *Critique of Pure Reason*, giving rise to a feeling of vertigo which stayed with him for the rest of his life.

---

[1] Originally published as "The Philosopher of the Physicists: The Legacy of Erhard Scheibe" by B. Falkenburg, *J. Gen. Philos. Sci.* 42 (2011), 1–15, together with the list of Scheibe's works underlying Appendix C.

E. Scheibe, *The Reduction of Physical Theories*, Fundamental Theories of Physics 207,
https://doi.org/10.1007/978-3-662-65000-4

He finished his secondary school education in Singen/Hohentwiel, in south-west Germany, receiving his university entrance diploma. In 1946, he began his studies in Göttingen, concentrating on mathematics, physics and philosophy. He soon belonged to the circle of young scholars gathered around Werner Heisenberg and Carl-Friedrich von Weizsäcker. In 1952, he met his future wife, Maria Elgert-Eggers, at that time still a schoolgirl, giving her private lessons in mathematics and physics, and also playing four-handed piano with her. He received his doctorate with a thesis in mathematics (1955),[2] obtained a fellowship at the Max Planck Institute for Physics in Göttingen, working with Heisenberg, and then, two years later, proceeded to Hamburg as von Weizsäcker's Assistant. In 1958, he and Maria married. Within a few years, they had three children, and from this point on, Erhard Scheibe's personal life was as unspectacular as his academic career became outstanding.

II.

Scheibe obtained his Habilitation in 1963 in Hamburg with a philosophical study of quantum mechanics. His inaugural lecture on the philosophy of Plato was accepted by the renowned journal for ancient philosophy, Phronesis (1967a), demonstrating even at this early stage his remarkable philosophical range.

The Habilitation Thesis, entitled *Die kontingenten Aussagen in der Physik (The Contingent Propositions of Physics)*, was published a year later (1964) with a preface by von Weizsäcker. The 'contingent' propositions of physical theories are the statements about physical systems which are not fixed by the laws of physics—in particular, initial values, boundary conditions, etc. By focusing on them, the book elucidates the similarities and distinctions between the classical and the quantum descriptions of a single physical system. This approach is in contrast to the various attempts at formulating a 'quantum logic', which were initiated by Johann von Neumann and Garrett Birkhoff in the 1930s and further elaborated by Peter Mittelstaedt in von Weizsäcker's circle. Scheibe's approach aims to show how quantum theory may safely be established on the grounds of classical logic, with recourse to the Copenhagen interpretation of quantum theory, but without any need of quantum logic. The approach follows Bohr's lines of thought, according to which the language of classical physics and the structure of quantum theory are in a certain sense complementary. Scheibe interprets the relation between classical and quantum concepts in epistemological terms, emphasizing that quantum theory cannot be understood without classical physics, while on the other hand quantum theory also contributes to a better understanding of the foundations of classical physics.

Methodologically, the book follows the mathematical concept of a theory structure developed by Günther Ludwig along the lines set out by von Neumann. This is in decided contrast to von Weizsäcker's much more informal philosophical work. Scheibe's life-long project was to bring as much mathematical clarity as possible into

[2] All quotations of Scheibe's works refer to the list given in Appendix C.

the philosophy of physics, and in particular, into the conceptual incompatibilities in the foundations of physics. A basic conceptual distinction from 1964 concerns the structure and the evolution or behavior of a physical system. This distinction gives rise to the following trisection of the propositions of a system into:

(1) (nomological) propositions about structure, which describe the state space;
(2) nomological propositions about behavior, which describe the dynamic evolution;
(3) contingent propositions about behavior, which describe the actual physical state.

The main differences between quantum theory and the theories of classical physics are reflected in the contingent propositions, as the book shows in great detail. This is closely related to the point that classical theories are deterministic whereas quantum theory is not. The quantum indeterminism is due to a specific epistemic character of the contingent propositions in quantum theory. And it is just this specific epistemic character which allows us to avoid quantum logic. Indeed, the main result of the book is that there are only two options for interpreting the standard version of quantum theory: either an ontological version of the theory built on the foundation of non-classical quantum logic, or an epistemic version, built this time on classical logic.

In 1964, as well as publishing this influential book, Scheibe also received a call to Göttingen as a full Professor of Philosophy. Three years later, he turned down a call which would have taken him back to Hamburg. Instead, he and his family would remain in Göttingen for 18 years, where he occupied himself above all with the structure of physical theories.

In the first decade of this period, he published *The Logical Analysis of Quantum Mechanics* (1973). The book continues the project of the Habilitation Thesis, that is, to reconstruct quantum theory and to compare it to classical mechanics. Its main task is to clarify what has been called the *Copenhagen interpretation* or 'orthodox' version of quantum theory. The introduction stresses how difficult this task turned out to be, given that the quantum 'orthodoxy' emerged in several versions from the work of Bohr, Heisenberg, von Neumann, Wolfgang Pauli, and others. Bohr's informal approach, which did not give rise to any rigorous version of quantum mechanics, differs substantially from the mathematical approaches of Heisenberg and von Neumann. Scheibe emphasizes that reconstructing the content of the latter in axiomatic terms does not give rise to a single unified theory either (1973, 5). For this reason, he took the decision to reconstruct only the most important fragments of quantum theory as well as their counterparts in classical mechanics and classical statistical mechanics.

The book begins with a crystal clear presentation of Bohr's philosophy of complementarity. Here, the aim is to clarify the arguments that underlie Bohr's various definitions of complementarity and the development of his thought over the two decades of the Bohr-Einstein debate. The remaining chapters give axiomatic reconstructions of quantum mechanics in Hilbert space in terms of quantities (or observables) and in terms of algebraic properties; the deterministic and indeterministic features of classical and quantum mechanics; the von Neumann programme; the reduction of states and the measurement process; and the problem of hidden parameters. The book ends with a formal analysis of the thought experiment of Einstein, Podolsky and Rosen (EPR), the high point of the Bohr-Einstein debate. In search of logical clarity, Scheibe analyses the reality criterion and the completeness conditions of the EPR paper by way of an extension of ordinary quantum mechanics that gives rise to a logical contradiction.

Edward MacKinnon's detailed review emphasizes that the book, "essentially a defense of the adequacy and logical consistency of the orthodox Copenhagen interpretation of quantum mechanics, sets a new standard for rigorous logical analysis of this difficult topic" (McKinnon 1975, 96). In fact, the book shows that there is no easy route to the establishment of the unity of physics. On the one hand, there is the gap between the classical theories and quantum mechanics. On the other hand, even the attempt to give a logically rigorous formulation of 'the' quantum theory behind the standard textbook accounts ends up in a variety of axiomatic fragments that can by no means be unified in a simple way. In the following decades, Scheibe would be engaged with just this apparent fragmentation of physical theory, and what it means for the unity of physics. With his patient efforts to work out how, despite this, unity might nevertheless be established, he pursued the life-long goal of his teacher, von Weizsäcker.[3]

III.

In addition to the books, there were numerous lectures and essays covering themes from the concept of cause to scientific explanation, from Leibniz to Kant, from the laws of physics to the laws of nature, from the structure of physical theories to the progress of physics, not to mention the many technical articles on quantum mechanics and special or general relativity. Scheibe's reputation grew. He became, not just one of the most important philosophers of the exact sciences, but the one who shaped the philosophy of physics in the German speaking world. In 1975, he joined the Editorial Board of this journal. In 1977, he became a full member of the Academy of the Sciences of Göttingen. In the academic year 1979–80, he was a fellow at the Center for Interdisciplinary Research (ZIF) at Bielefeld and at the Center for Philosophy of Science in Pittsburgh. In 1982, he became a corresponding member of the Academy of Science and Letters of Mainz, and of the Académie Internationale de la Philosophie des Sciences in Brussels.

From 1982 until his retirement and nomination as Emeritus Professor in 1992, Scheibe was the holder of a then newly-founded Chair of Philosophy at Heidelberg, one dedicated to philosophical issues concerning the formal and natural sciences. His move to Heidelberg marked a shift in his interests from the formal structure of physical theories to more general features of the philosophy of physics. Once established in Heidelberg, he turned his attention more to the historical and conceptual background of physics, including the ways in which physics is rooted in the traditions

[3] Von Weizsäcker died in 2007. See the obituary Drieschner 2008 and the bibliography Anacker and Schöttler 2008.

of rationalism and empiricism. The shift of interest was in particular marked by the topic of his inaugural lecture Kohärenz und Kontingenz. Die Einheit der Physik und die rationalistische Tradition (Coherence and Contingency. The Unity of Physics and the Tradition of Rationalism) (1986c). In both research and teaching, he sought to investigate the characteristic ways the physics of the 20th century had combined both traditions. In early 1987 he spent three months as an invited professor at the University of California in Irvine. In the academic year 1987/88 he was a fellow at the Wissenschaftskolleg (Institute for Advanced Studies) in Berlin. In 1988, he held the Presidency of the Leibniz Committee of the Academy in Göttingen. From 1989 to 2002, he was a member of the Editorial Board of the journal *Philosophia Naturalis*.

In addition to his outstanding work on the axiomatic foundations and the philosophical background of modern physics, Erhard Scheibe was an influential academic teacher. He exuded a unique combination of scientific rigor and precision, combined with warmth and friendship, a combination which embraced and inspired both his students and the Humboldt Fellows and other scientific guests from all over the world who came to work with him. In the first decade of the Göttingen years, he supervised the Habilitation of Lorenz Krüger, who later became a full professor in Bielefeld and Göttingen, and, in the years before his untimely death, founded the Max Planck Institute for the History of Science in Berlin. Other by now well-established students and Assistenten from the Göttingen or Heidelberg years were Werner Diederich, Andreas Hüttemann, Bernd-Olaf Küppers, Ulrich Majer, and Manfred Stöckler, as well as the author of this article.

But even though many of them were, in time, appointed to chairs in philosophy of science at German universities, there was never a 'Scheibe school' as such, and Scheibe was not the least bit interested in the indoctrination of anyone. Indeed, owing to his modest nature, his respect for the individual approaches of others, and also his need for solitary work in view of the difficulties of his projects, Scheibe has no direct followers. Having the highest esteem for scientific objectivity, he considered himself to be a 'Diener der Wissenschaft', a true servant of science. His reference group was made up from colleagues in physics and mathematics, with whom he was involved in many discussions. In particular, he maintained an extensive correspondence on the foundations of quantum mechanics with Günther Ludwig and Hans Dieter Zeh.

The lack of a Scheibe school stood in striking contrast to the Erlangen and Stegmüller schools of that time. Scheibe repeatedly debated with the members of the latter, in particular Carlos-Ulises Moulines, over the formal tools to be employed in reconstructing the architectonics of physics. Outside the German speaking world, Scheibe was, perhaps, not as generally well-known as the significance and depth of his work merited. But experts knew, a fact which is testified to by the frequency with which he was invited to international conferences, and also by the number of young scholars who came to study with him under the auspices of the Humboldt Stiftung.

It is a sign of Scheibe's engagement as a teacher, but possibly also of the worsening situation as regards German university research, that he was able to complete his central work only with the freedom which retirement brings. After his retirement in 1992, he and Maria moved back to Hamburg. The book *Die Reduktion physikalischer*

Theorien. Ein Beitrag zur Einheit der Physik (The Reduction of Physical Theories. A Contribution to the Unity of Physics) appeared in two volumes in 1997 and 1999. It is a monumental investigation of the unity of physics in the face of apparently incommensurable theories. Whoever is to profit from a study of this must bring to it a sound knowledge of the mathematical and logical foundations of modern theoretical physics. Springer is currently preparing an Engl. transl. of this seminal work.

Bearing as it does the fruits of Scheibe's profound work on the axiomatic foundations of physics, the book is his answer to the philosophical discussion surrounding incommensurability. From his insights into the axiomatic fragments of physical theories, he knew that the unity of physics is not available in any cheap or easy way, and he explained it in terms of inter-theoretic relations. This approach gave rise to a completely novel account of reduction and to a new understanding of what progress in physics amounts to. The crucial new idea is that there is a unique way that one physical theory can be reduced to another, but (in general) many. These have to be spelled out in terms of different kinds of inter-theoretic reduction relations, in particular, various kinds of exact reduction (equivalence, embedding, refinement, extension, unification), approximate reduction and partial reduction. Scheibe emphasizes that, in contradistinction to all previous work on reduction, his new approach is *synthetic* or *recursive*; it is not centered around a general pre-determined concept of theory reduction, but rather aims at finding several non-decomposable, or elementary, kinds of reduction which are as specific as possible (1997, 3-4). The resulting concepts of reduction and of unity are constructive, multi-faceted, and open to extensions.

IV.

The first volume of the *Reduction* (1997) develops the theory. The second volume (1999) applies it in extensive case studies to the conceptual caesura within the foundations of physics. Since the virtues and the main results of these investigations have been the subject of several detailed reviews (see Appendix C), here it is perhaps of more interest to sketch how their topics are interwoven with the rest of Scheibe's work, in particular with the numerous articles which emerged from Scheibe's years in Göttingen and Heidelberg.

The inter-theoretic relations between classical mechanics, classical statistical mechanics, and quantum theory puzzled Scheibe for many years, beginning with the first two books (1964, 1973). From 1973 on, he took up the topic of theory reduction explicitly, in concrete case studies (1973b, 1981a, 1988h, 1997d), in articles on approximate explanation (1973c), the comparison of theories (1975, 1976bd, 1982b, 1986b), and and the explanation of one theory by another (1976ae). From 1984 on, he linked these topics to the problem of progress in physics (1984b), deepening the latter in subsequent articles (1985c, 1988a). The last-mentioned articles lay the foundation for the first chapter of Die Reduktion, which discusses the physicists' and the philosophers' views about theoretical progress (1997, 13-34). These views indeed turn out to be much more subtle than is generally allowed for within the philosophy of science: the progress of physics does not preclude there being some highly significant conceptual breaks; scientific revolutions do not preclude the existence of some standards for the comparison of theories; and, on a closer look, neither the physicists nor the philosophers deny this.

The detailed formal explanations of the book are based on a series of articles on the structure of physical theories, the different ways of reconstructing them, and the relation between mathematics and the axiomatic approach to physics (1979, 1981c/82a, 1983b, 1986a, 1988c). They compare Ludwig's approach to that of the Stegmüller school, which in turn is based on the model-theoretic approach of Joseph D. Sneed, and they begin to shape the way in which Scheibe's own views differ. He insists on a distinction between mathematical and physical concepts, which is not taken into account by the empiricist distinction of observables and unobservables. According to the second chapter of em Die Reduktion, the structure of a physical theory indeed rests on three kind of objects, namely mathematical, physical, and mixed objects (1997, 63–75). To the physical objects belong in particular *theoretical entities* such as the strength of an electromagnetic field. And this feature shows that Scheibe's approach lies somewhere in between rationalism and empiricism, just as does 'the' philosophy of the physicists which he started to investigate in a systematic way in Heidelberg. (See the books 2001 and 2006/2007.)

Some of the earlier articles indicate in addition that the resulting theory of reduction would deal with various different kinds of inter-theoretic relations, such as approximative explanation (1973c), different types of successor relations between theories (1983a), and the distinction between equivalence and reduction (1988h). But the many different variants of exact, approximate, and partial reduction discussed in chapters IV–VI of Die Reduktion, and the many challenging cases provided by the theories of relativity, the notion of micro-reduction, and the reduction of classical to quantum cases set out in chapters VII–X (1999), could only be worked out in detail after Scheibe's retirement.

Let me conclude my remarks on *Die Reduktion* with a look at some philosophical aspects of the unity of physics advanced there. How does it relate to traditional views? For most physicists, the unity of physics is a heuristic principle of theory construction which is based on faith in a corresponding structural unity of nature. Newton, Maxwell, and Einstein used this principle in order to construct more and more comprehensive theories. Planck explained the unity of physics in the rationalist, if not Platonist, terms of an all-embracing mathematical structure behind the phenomena, sketching a 20th century version of Galileo's dignified book of nature. In contrast to this rationalist faith, Kant stressed in the Critique of Judgment that the idea of unity of nature does not have any certain foundation in the objective structure of the phenomena. According to him, the unity of nature is nothing but a regulative principle which governs our own search for systematic, conceptually unified knowledge.

Following Kant and Bohr, Scheibe's teacher von Weizsäcker had tried to make the bridge between both views, in a *neo-transcendental* approach to physics, and to the Copenhagen interpretation of quantum theory in particular. On the occasion of von Weizsäcker's 80th birthday, Scheibe discussed this approach and its relation to the goal of a 'final' theory of physics. According to von Weizsäcker, such a final theory of physics should no longer be built on empirical laws but only on the conditions of possible experience (1993c). In his discussion, Scheibe emphasizes that this neo-transcendental approach comes along with an extreme reductionism, which, however, in some aspects comes close to Thomas S. Kuhn's or Paul Feyerabend's views about theory change.

The article appeared in the same year as the first sketch of the new theory of reduction (1993b). This theory and the resulting many-faceted account of the unity of physics is in sharp contrast to von Weizsäcker's neo-transcendental approach, but it, too, bridges the gap between the traditional views mentioned above. Scheibe's unity of physics is synthetic or constructive, just as Kant had suggested. However, it is not a priori, as von Weizsäcker had assumed, but constructed from the empirical laws of physics. At the same time, this unity of physics is rational; it rests on the structure of the phenomena, just as the founders of modern physics had claimed. But it differs substantially from the rationalist idea of an all-embracing, final theory of physics. It is made up of many axiomatic fragments, which are linked by many kinds of inter-theoretical relations. According to Scheibe, the unity of physics is piecemeal, though it is surely not piecemeal in Nancy Cartwright's understanding of that term.

In a certain sense, Scheibe's approach stands in the long-neglected Neo-Kantian tradition of philosophy of science, which was established by the Marburg school of Hermann Cohen, Paul Natorp (1910, 18), and Ernst Cassirer. The affinity lies mainly in one feature. It was Natorp who first emphasized that physical reality is "nicht gegeben, sondern aufgegeben", that is, not given but posed for us as a problem, a problem which has to be resolved by construction.[4] And it was Einstein (1949, 680) who took up this phrase (in the reply to his critics), by commenting that he was late in coming to appreciate Kant's philosophy.[5] Scheibe surely knew the latter source, but perhaps not the former. He would probably have listed the Neo-Kantians among the defenders of apriorism, like Kant himself and von Weizsäcker.

The constructivist aspects of the Neo-Kantian approach are not to be confused with more recent versions of constructivism. In his article *Mißverstandene Naturwissenschaft (Science Misunderstood)* (1979b) sharply criticizes several recent constructivist views, such as the Erlangen school, the radical constructivism advanced by neuroscientists such as Heinz von Foerster, or the social constructivism propagated by Andrew Pickering from the Edinburgh school, Bruno Latour, and other post-modern thinkers. These views differ substantially from the Neo-Kantian con-

[4] "So kann also von keinem 'gegebenen' Gegenstand mehr die Rede sein; also auch nicht von Erkenntnis als bloßer Analyse dieses Gegebenen. Gerade der Gegenstand vielmehr ist *Aufgabe*, ist *Problem* ins Unendliche." Matters that are "gegeben", i.e. , given, or "aufgegeben", i.e. , posed as problems, were the starting points respectively of the traditional 'analytic' and 'synthetic' methods. These were based in ancient geometry, used by the rationalist philosophers in their accounts of an ars inveniendi, and were still well-known to the Neo-Kantian philosophers around 1900. Scheibe discussed these methods in his Heidelberg seminars on the classical texts of rationalism.

[5] "I did not grow up in the Kantian tradition, but came to understand the truly valuable which is to be found in his doctrine, alongside of errors which today are quite obvious, only quite late. It is contained in the sentence: 'The real is not given to us, but put to us (aufgegeben) (by way of a riddle).'"

structivist approaches, among which, I would suggest, we place Scheibe's theory of reduction. The common ground of Scheibe's and Natorp's views is the insight that the unity of physics, and with it our concept of physical reality, is still under construction, and will perhaps forever be.

V.

A collection in English of Scheibe's most important papers and lectures followed, making a great deal more of his work accessible to the English speaking philosophical world. *Between Rationalism and Empiricism: Selected Papers in the Philosophy of Physics* (2001) has eight parts. Each of them is preceded by a brief introduction by the author. The first part, *Between Rationalism and Empiricism*, is programmatic, as is, in a sense, the whole book. This has as its theme the sophisticated philosophical attitudes of the most prominent physicists, 'wavering' as it were between these two philosophical extremes. Scheibe points out that these extremes themselves do not help to understand the structure of physics. It is worth quoting a lengthy passage from the introduction, which makes explicit the core of his philosophical convictions:

> Einstein has gone so far as to call the physicist an 'unscrupulous opportunist' who, depending on the circumstances, appears as a realist or an idealist or a positivist or even a Platonist ...But this wavering attitude is by no means the product of opportunism. Rather it should be viewed as protest against exaggerated positions of philosophers of science which bring their inventors repute and even fame but lead them astray from the road of to truth. ...But whoever sets out ...to look for a workable synthesis is not an opportunist. He is not even an eclecticist in the pejorative sense of the word which we have in mind whenever we take the extreme positions to be the genuine ones. This is not justified if every attempt at their application to a further developed science fails. We have to turn the tables and view our sciences as the empirical material for an appropriate theory of scientific knowledge. A thoroughgoing analysis has to be started before one comes to generalizations. (Scheibe 2001, 1)

Here we see the deepest reason for the fact that there is no Scheibe school. According to Scheibe, the only task of the philosophy of science is to analyze the structure of scientific theories, the ways in which they relate to each other, the reasons which led their creators to establish them, and the views of the scientists about the way in which the theories express physical reality. Any philosophical doctrine that might give rise to a school would be contrary to this spirit. It follows that Scheibe was much more inclined to give support to students who were investigating philosophical problems connected to actual scientific practice, as opposed to investigating what he used to call the "self-generated problems of philosophy of science".

For the same reason, he was increasingly convinced that the structure and the conceptual revolutions of physics can be properly understood only by combining historical and systematical studies. This is clearly seen from the choice of articles he selected for inclusion in the volume. Part I of the collection is centered around a paper on Kant's apriorism (1988e), the article on Weizsäcker's neo-transcendental approach mentioned above (1993c), and the evening lecture from the Stuttgarter

Hegel-Kongress 1993 (1994a), which gives the title to the collection as well as to its first section.

Part II of the collection, entitled *The Philosophy of the Physicists*, contains papers that lay the foundation for the project from which Scheibe's last book (2006) emerged. The article on the physicists' conception of progress (1988a) overlaps with the first chapter of *Die Reduktion physikalischer Theorien.* The paper on Einstein's epistemology (1992b), given in Heidelberg as his farewell lecture, explains the details of Einstein's "wavering attitude" between rationalism and empiricism. The investigation of Heisenberg's concept of a closed theory (1993a), in turn, is related to Scheibe's specific approach to reconstructing the structure of physical theories.

The papers of Part III of the collection, entitled *Reconstruction*, defend the enterprise of a 'rational' reconstruction of physical theories (1984a, 1988b). This notion was in particular attacked by Feyerabend, and there is a sense in which it might be seen as in conflict with the ideal of focusing on the practice of physics. In addition, the papers here demonstrate the ways in which Scheibe's approach differs from Ludwig's or Sneed's (1979, 1981c/1982a).

Part IV of the collection, *Laws of Nature*, contains articles that extend the ideas on coherence and contingency (1989a), first presented in the inaugural lecture given in Heidelberg in 1982, to observations about physical laws and the way to express them (1991a), the reference of quantum mechanics to physical systems (1991f), a tension between the generality of laws and the uniqueness of the universe (1991b), and the limitations of knowledge which may arise from these problems (1998a). The section stands on its own, not only in Scheibe's work, but also in the philosophy of science in general. According to the review by Maarten Van Dyck, it takes up the "perennial question about Laws of Nature" in terms of the "tension that exists between on the one hand and their content, which [...] concerns individual systems, and on the other hand their form, which (in)famously is to be universal", a tension which is closely related to the "basic opposition between empiricism and rationalism" (van Dyck 2000, 104). But the question remains open of what exactly this means for the validity of physical laws and the claims of scientific realism.

The ideas underlying these articles may have been influenced by many years of discussions with Nancy Cartwright. He first met her at the ZIF in Bielefeld, then at the Wissenschaftskolleg in Berlin, and at many conferences. The common elements between Erhard and Nancy were the convictions that the best strategy in philosophy of science is to study scientific practice, that the structure of physics is piecemeal, and that this fragmentary structure is closely related to certain limitations in our physical knowledge. The point of departure between them was the question of whether it is worthwhile or not to make the effort to give a rational reconstruction of the axiomatic fragments and their inter-theoretic relations. The other parts of Scheibe's collection put together further articles on the topics of Scheibe's books. Part V, *Reduction*, contains several of the papers mentioned in the last section. Part VI, *Foundations of Quantum Mechanics*, incorporates a more popular paper on the Copenhagen interpretation (1990a) and five highly technical papers on quantum logic, hidden variables, EPR, and Bell's theorems published between 1985 and 1993. Part VII is called *Spacetime, Invariance, Covariance*. Here, a paper on Herrmann Weyl (1988f) is brought together with three papers on the subtle distinctions of invariance and covariance, which were written between 1982 and 1994. Part VIII, *Mathematics and Physics*, unites papers from two decades concerned with the relationships between mathematics and physics, papers on Kant (1977), Leibniz (1990b), and the scope of mathematics in physics (1986a, 1997a).

VI.

In 2003, Erhard Scheibe was made an honorary member of the Gesellschaft für Analytische Philosophie (GAP). By then, he was no longer able to travel and to participate in the ceremony. Indeed, his final book, *Die Philosophie der Physiker (The Philosophy of the Physicists)* (2006, 2007), was wrested from the ravages of an increasingly debilitating illness, and Scheibe passed away on January 7, 2010, in Hamburg. The presentation given by this final work of the philosophical convictions of the most important physicists of the 20th century became an enormous success. It is now to be read, it seems, as an appropriate legacy of Erhard Scheibe.

Acknowledgements I would like to thank Raphael Bolinger for preparing the bibliography and Michael Hallett for improving my English.

References

Anacker, M., and Schöttler, T. 2008. Carl Friedrich von Weizsäcker-Bibliographie. *J. Gen. Philos. Sci.* 39, 179–244.

Drieschner, M. 2008. Carl Friedrich von Weizsäcker. *J. Gen. Philos. Sci.* 39, 1–16.

Einstein, A. 1949. Reply to my Critics, In: P. A. Schilpp (Ed.), *Albert Einstein: Philosopher–Scientist*. La Salle: Open Court, 663–688.

MacKinnon, E. 1975. In: *Philosophy of Science* 42, No. 1, 96–100.

Natorp, P. 1910. *Die logischen Grundlagen der exakten Wissenschaften*. Leipzig und Berlin: Teubner. Zitiert nach der Zweiten, durchgesehenen Auflage, Leipzig und Berlin: Teubner 1921.

Planck, M. 1908. Die Einheit des physikalischen Weltbilds. In: *Vorträge und Erinnerungen*. Stuttgart 1949, 28–51.

van Dyck, M. 2000. In: *Philosophica* (Belgium) 65, 101–106.

Appendix C
Erhard Scheibe's Works

The list of Scheibe's works is taken from *J. Gen. Philos. Sci.* 42 (2011), 10–15.[6]

Books

1964 *Die kontingenten Aussagen in der Physik. Axiomatische Untersuchungen zur Ontologie der klassischen Physik und der Quantentheorie.* Mit einem Vorwort von C. F. v. Weizsäcker. Frankfurt a. M.: Athenäum.

1973 *The Logical Analysis of Quantum Mechanics.* Translated by J. B. Sykes. Oxford: Pergamon Press.

1997 *Die Reduktion physikalischer Theorien. Ein Beitrag zur Einheit der Physik. Teil I: Grundlagen und elementare Theorie.* Berlin: Springer.

1999 *Die Reduktion physikalischer Theorien. Ein Beitrag zur Einheit der Physik. Teil II: Inkommensurabilität und Grenzfallreduktion.* Berlin: Springer.

2001 *Between Rationalism and Empiricism. Selected Papers in the Philosophy of Physics.* Ed. by B. Falkenburg. Berlin: Springer.

2006 *Die Philosophie der Physiker.* München: C. H. Beck. Revised paperback edition: 2007.

Editions

1973a *Einheit und Vielheit. Festschrift für C. F. v. Weizsäcker zum 60. Geburtstag* (with G. Süßmann). Göttingen: Vandenhoeck & Ruprecht.

[6] Cf. Note 1 above. Some errors have been corrected, and for reasons of compatibility with the present book the order of a few references has been modified.

E. Scheibe, *The Reduction of Physical Theories*, Fundamental Theories of Physics 207,
https://doi.org/10.1007/978-3-662-65000-4

1977a *Logik, Ethik, Theorie der Geisteswissenschaften. XI. Deutscher Kongreß für Philosophie. Göttingen 5. -9. 1975* (with G. Patzig and W. Wieland]. Hamburg: Meiner.

1988a *The Role of Experience in Science. Conf. Acad. Int. Phil. Sci., Heidelberg 1986*. Berlin: de Gruyter 1988.

Articles and Reviews

1952 Über einen verallgemeinerten affinen Zusammenhang. *Mathematische Zeitschrift* 57, 65–74.

1957 Über das Weylsche Raumproblem (Dissertation). *Journal für die reine und angewandte Mathematik* 197, 162–207.

1958 Komplementarität und Logik III. Mehrfache Quantelung (with C. F. v. Weizsäcker and G. Süßmann). *Zeitschrift für Naturforschung* 13a, 705–721.

1960 Über hermitische Formen in topologischen Vektorräumen 1. *Annales Academiae Scientiarum Fennicae Mathematica*. Heft 294, 1–30.

1962 Zum Problem der maschinellen Translation von Sprachen. In: *Die Kunst der Translation. Jahrbuch Bayrische Akademie der Künste 8*, 186–205.

1967a Über Relativbegriffe in der Philosophie Platons. *Phronesis* 12, 28–49 (Inaugural lecture Hamburg].

1967b Die Einheit der Zeit. In: E. Fries (Ed.), *Festschrift für Joseph Klein zum 70. Geburtstag*. Göttingen: Vandenhoeck & Ruprecht 1967, 53–72.

1967c Über das Verhältnis von Wissenschaft und Philosophie. *Neue Sammlung* 7, 296–309.

1967d Zum Problem der Sprachabhängigkeit in der Physik. In: H. -G. Gadamer (Ed.), *Das Problem der Sprache. VIII. Deutscher Kongress für Philosophie: Heidelberg 1966*. München: Fink, 313–334.

1967e Rezension: G. Patzig: Die aristotelische Syllogistik. 2. Auflage. Göttingen: Vandenhoeck & Ruprecht (erste Auflage 1963). *Gnomon* 39, 454–464.

1968a Bibliographie zu Grundlagenfragen der Quantenmechanik. *Philosophia naturalis* 10, 249–298.

1968b Philosophie und Physik. Veröffentlichungen in deutscher Sprache. In: R. Klibansky (Ed.), *Contemporary Philosophy. A Survey*. Florenz: La Nuova Italia Editrice. Vol. II, 200–217.

1969 Bemerkungen über den Begriff der Ursache. In: H. H. Holz und J. Schickel (Eds.), *Vom Geist der Naturwissenschaft*. Zürich: Rhein, 105–134. Engl. transl.: Remarks on the Concept of Cause. In: D. E. Christensen et al. (Eds.), *Contemporary German Philosophy*. Vol. 4. London: Penn State University Press 1984. 223–243. Repr. in: Scheibe 2001, 4–22.

1970 Ursache und Erklärung. In: L. Krüger (Ed.), *Erkenntnisprobleme der Naturwissenschaften. Texte zur Einführung in die Philosophie der Wissenschaft*. Köln: Kiepenheuer & Witsch, 253–275.

1971 Ein vernachlässigter Aspekt physikalischer Erklärung. *Naturwissenschaften* 58, 1–6.
1972 Der Zeitbegriff der Physik. Ein Beitrag zum Gespräch zwischen den Natur und Geisteswissenschaften. *Saeculum* 23, 236–251.
1973b Die Erklärung der Keplerschen Gesetze durch Newtons Gravitationsgesetz. In: E. Scheibe and G. Süßmann (Eds.), *Einheit und Vielheit. Festschrift für C. F. v. Weizsäcker zum 60. Geburtstag*. Göttingen: Vandenhoeck & Ruprecht, 98–118. Engl. transl.: The Explanation of Kepler's Laws, in: Scheibe 2001, 306–323.
1973c The Approximative Explanation and the Development of Physics. In:P. Suppes et al. (Eds.), *Logic, Methodology and Philosophy of Science IV*. Amsterdam: North Holland, 930–942.
1973d Gibt es eine moderne Naturphilosophie? In: M. Ebhardt (Ed.), *Reflexe—Begegnungen. Eduard Meyer zum 85. Geburtstag*. Bremen: Schünemann, 55–70 (Inaugural lecture Göttingen).
1973e Wissenschaft und Wahrheit. *Gymnasium* 80 56–77.
1973f Gesetzlichkeit und Kontingenz in der Physik. In: K. Hübner und A. Menne (Eds.), *Natur und Geschichte. X. Deutscher Kongress für Philosophie*. Kiel 1972. Hamburg: Meiner, 170–189.
1974 Popper and Quantum Logic. *British Journal for the Philosophy of Science* 25, 319–328.
1975 Vergleichbarkeit, Widerspruch und Erklärung. In: R. Haller and J. Götschl (Eds.), *Philosophie und Physik*. Braunschweig: Vieweg, 57–71.
1976a Gibt es Erklärungen von Theorien? *Allg. Zeitschr. f. Philosophie* 1, 26–45. Engl. transl.: Are there Explanations of Theories? In: D. E. Christensen et al. (Eds.), *Contemporary German Philosophy*, Vol. 3. London: Penn State University Press 1983, 141–158. Repr. in: Scheibe 2001, 324–338.
1976b Conditions of Progress and the Comparability of Theories. In: R. P. Cohen et al. (Eds.), *Essays in Memory of Imre Lakatos*. Dordrecht: Reidel, 547–568.
1976c Kausalgesetz. In: K. Gründer (Ed.), *Historisches Wörterbuch der Philosophie*. Vol. 4. Basel: Birkhäuser, 789–798.
1976d Zur Logik des Theorienvergleichs. In: B. Kanitscheider (Ed.), *Sprache und Erkenntnis. Festschrift für G. Frey zum 60. Geburtstag*. Innsbruck: Inst. f. Sprachwissenschaft d. Univ. Innsbruck, 13–25.
1976e Wie beurteilt man eine physikalische Theorie? *Georgia Augusta* 25 19–29.
1977 Kants Philosophie der Mathematik. *Mittlgn. Math. Ges. Hamburg* X, 353–372. Engl. transl. in: Scheibe 2001, 517–534.
1978 Der naturwissenschaftliche Prozessbegriff. In: K. G. Faber und C. Meier (Eds.), *Theorie der Geschichte*, Vol. 2. München: DTV, 374–412.
1979 On the Structure of Physical Theories. In: I. Niiniluoto und R. Tuomela (Eds.), *The Logic and Epistemology of Scientific Change*. Amsterdam: North Holland, 205–224. Repr. in: Scheibe 2001, 160–174. German translation: Über die Struktur physikalischer Theorien. In: W. Balzer und M. Heidelberger (Eds.), *Zur Logik empirischer Theorien*. Berlin: de Gruyter 1983, 169–188.

1980 Wissenschaftstheoretische Probleme der Physik. In: J. Speck (Ed.), *Handbuch wissenschaftstheoretischer Begriffe*. Göttingen: Vandenhoeck & Ruprecht, 470–480.

1981a Eine Fallstudie zur Grenzfallbeziehung in der Quantenmechanik. In: J. Nitsch et al. (Eds.), *Grundlagenprobleme der modernen Physik. Festschrift für P. Mittelstaedt zum 50. Geburtstag*. Mannheim: Bibliographisches Institut 257–269. Engl. transl.: A Case Study Concerning the Limiting Case Relation in Quantum Mechanics. In: Scheibe 2001, 339–351.

1981b Quantentheorie und verborgene Parameter. Eine logische Analyse. *Physikunterricht* 15, 56–74.

1981c A Comparison of Two Recent Views on Theories. In: A. Hartkämper und H. -J. Schmidt (Eds.), *Structure and Approximation in Physical Theories*. New York: Plenum, 197–215.

1982a A Comparison of Two Recent Views on Theories. *Metamedicine* 3, 233–253 (slightly extended version of 1981c]. Repr. in: Scheibe 2001, 175–194.

1982b Zum Theorienvergleich in der Physik. In: K. M. Meyer-Abich (Ed.), *Physik, Philosophie und Politik. Festschrift für C. F. v. Weizsäcker zum 70. Geburtstag*. München: Hanser, 291–309.

1982c Invariance and Covariance. In: J. Agassi und R. P. Cohen (Eds.), *Scientific Philosophy Today. Essays in Honour of Mario Bunge*. Dordrecht: Reidel 311–331. Repr. in: D. Mayr und G. Süßmann (Eds.), *Space, Time, and Mechanics. Basic Structures of a Physical Theory*. Dordrecht: Reidel 1983, 125–147. Repr. in: Scheibe 2001, 457–474.

1983a Two Types of Successor Relations Between Theories. In: *Zeitschrift für allgemeine Wissenschaftstheorie* XIV, 68–80.

1983b Ein Vergleich der Theoriebegriffe von Sneed und Ludwig. In: P. Weingartner und H. Czermak (Eds.), *Erkenntnis- und Wissenschaftstheorie. 7. Internationales Wittgenstein Symposium Kirchberg/Wechsel 1982*. Wien: Hölder-Pichler-Tempski, 371–383.

1983c Kriterien zur Beurteilung der Naturwissenschaften. In: *Wissenschaft und Wissenschaftsbegriff. Abhandlungen der Akademie der Wissenschaften und Literatur Mainz. Mathematisch-naturwissenschaftliche Klasse* Nr. 3, 25–41.

1984a Zur Rehabilitierung des Rekonstruktionisms. In: H. Schnädelbach (Ed.), *Rationalität. Philosophische Beiträge*. Frankfurt a. M.: Suhrkamp, 94–116. Engl. transl.: Towards a Rehabilitation of Reconstructionism. In: Scheibe 2001, 195–211.

1984b Explanation of Theories and the Problem of Progress in Physics. In: W. Balzer et al. (Eds.), *Reduction in Science. Structure, Examples, Philosophical Problems*. Dordrecht: Reidel, 71–94.

1985a Quantum Logic and some Aspects of Logic in General. In: P. Mittelstaedt und E. W. Stachow (Eds.), *Recent Developments in Quantum Logic*. Mannheim: Bibliographisches Institut 115–128. Repr. in: Scheibe 2001, 383–390.

1985b Some Aspects of Theoreticity in Physics. In: *La Nature de la Vérité Scientifique. Coll. Acad. Int. Phil. Sci., Brussels 1981*. Ciaco: Louvain-la-Neuve, 67–81.

1985c Die jüngste wissenschaftstheoretische Diskussion um den Begriff des Fortschritts. In: *Evolution. Studium Generale Univ. Heidelberg WS 1981/2*. Heidelberg, 111–129.

1985d Die Zunahme des Kontingenten in der Wissenschaft. *Neue Hefte für Philosophie* 24/25, 1–13.

1986a Mathematics and Physical Axiomatization. In: J. Vuillemin (Ed.), *Mérites et Limites des Méthodes Logiques en Philosophie*. Coll. Int. Fondation Singer-Polignac, Paris 1984. Paris: Vrin, 251–277. Repr. in: Scheibe 2001, 535–552.

1986b The Comparison of Scientific Theories. *Interdisciplinary Science Reviews* 11, 148–152.

1986c Kohärenz und Kontingenz. Die Einheit der Physik und die rationalistische Tradition. *Zeitschrift für philosophische Forschung* 40, 321–336 (inaugural lecture, Heidelberg).

1986d What Kind of Hidden Variables are Excluded by Bell's Inequality? In: P. Weingartner et al. (Eds.), *Foundations of Physics. 7th International Congress of Logic, Methodology and Philosophy of Science, Physics Section*. Wien: Hölder-Pichler-Tempsky, 252–271. Repr. in: Scheibe 2001, 391–401.

1987a Ganzheitsaspekte in Philosophie und Wissenschaft. In: *Jahrbuch der Akademie der Wissenschaften*. Göttingen: Vandenhoeck & Ruprecht, 39–54. Engl. transl.: Aspects of Wholeness in Science and Philosophy. In: S. N. and Y. B. Tripathi (Eds.), *Medical Systems with a Holistic Approach*. Varanasi 1993, 13–29. Repr. in: Scheibe 2001, 23–35.

1987b The Increase of Contingencies in Science. *Epistemologia* X, 171–186.

1987c Gibt es eine Annäherung der Naturwissenschaften an die Geisteswissenschaften? J. Assmann und T. Hölscher (Eds.), *Universitas* 42, 5–18. Repr. in: *Kultur und Gedächtnis*. Frankfurt a. M.: Suhrkamp, 1988, 65–83.

1988a The Physicists' Conception of Progress. *Stud. in History and Philosophy of Science* 19, 141–159. Reprined in: Scheibe 2001, 90–107. German version: Der Fortschritt der Physik im Selbstverständnis der Physiker. In: W. Muschik und E. Scheibe (Eds.), *Philosophie Physik Wissenschaftsgeschichte*. Berlin: TU Berlin, 1988, 165–183. German version reprinted in: M. Kroß (Ed.), *Vom Verständnis der Natur*. Einstein Forum, Jahrbuch 2000. Berlin: Akademie Verlag, 127–147.

1988b Paul Feyerabend und die rationalen Rekonstruktionen. In: P. Hoyningen-Huene und G. Hirsch (Eds.), *Wozu Wissenschaftsphilosophie? Positionen und Fragen zur gegenwärtigen Wissenschaftsphilosophie*. Berlin: de Gruyter, 149–171. Engl. transl.: Paul Feyerabend and Rational Reconstructions. In: Scheibe 2001, 212–227.

1988c Struktur und Theorie in der Physik. In: J. Audretsch und K. Mainzer (Eds.), *Philosophie und Physik der Raum-Zeit*. Mannheim: Bibliographisches Institut, 103–119.

1988d Do We Want to Be Irrational? Prof Scheibe replies to Prof. Hall. *Interdisciplinary Science Reviews* 13, 84–87.

1988e Kant's Apriorism and some Modern Positions. In: E. Scheibe (Ed.), *The Role of Experience in Science. Conf. Acad. Int. Phil. Sci., Heidelberg 1986*. Berlin: de Gruyter, 1–22. Repr. in: Scheibe 2001, 36–53.

1988f Hermann Weyl and the Nature of Spacetime. In:W. Deppert et al. (Eds.), *Exakte Wissenschaften und ihre philosophische Grundlegung. Internationaler Hermann Weyl Kongress Kiel 1985*. Frankfurt a. M.: Peter Lang, 61–82. Repr. in: Scheibe 2001, 475–489.

1988g Kovarianz und Gleichberechtigung. Bemerkungen zu einem Prinzip der allgemeinen Relativitätstheorie. *Philosophia naturalis* 25, 110–122. English version: Covariance and the Non-Preference of Coordinate Systems. In: G. G. Brittan (Ed.), *Causality, Method, and Modality. Essays in Honor of J. Vuillemin*. Dordrecht: Kluwer 1991, 23–40. Repr. in: Scheibe 2001, 490–500.

1988h Äquivalenz und Reduktion. Zur Frage ihres empirischen Status. *Conceptus* XII, 91–105.

1989a Coherence and Contingency. Two Neglected Aspects of Theory Successsion. *Nous* XXIII, 1–16. Repr. in: Scheibe 2001, 232–245.

1989b Die Kopenhagener Schule. In: G. Böhme (Ed.), *Klassiker der Naturphilosophie. Von den Vorsokratikern bis zur Kopenhagener Schule*. München: C. H. Beck, 374–392 and 436–438.

1989c Das Reduktionsproblem und die Einheit der Naturwissenschaften. In: P. and H. Czermak (Eds.), *Philosophie der Naturwissenschaften 13. Internationales Wittgenstein Symposium Kirchberg/Wechsel 1988*. Wien: Hölder-Pichler-Tempsky, 305–317.

1990a Die Kopenhagener Schule und ihre Gegner. In: J. Audretsch and K. Mainzer (Eds.), *Wieviele Leben hat Schrödingers Katze? Zur Physik und Philosophie der Quantenmechanik*. Mannheim: Bibliographisches Institut, 157–182. Engl. transl.: The Copenhagen School and Its Opponents In: Scheibe 2001, 402–418.

1990b Calculemus! Das Problem der Anwendung von Logik und Mathematik. In: I. Marchlewitz und A. Heinekamp. (Eds.), *Leibniz' Auseinandersetzung mit Vorgängern und Zeitgenossen*. Stuttgart: Steiner, 200–216. English version: Calculemus! The Problem of Application of Logic and Mathematics. *Knowledge Organization* 23 (1996), 67–76. Repr. in: Scheibe 2001, 553–570.

1991a Predication and Physical Law. *Topoi* 10 3-12. Repr. in: Scheibe 2001, 246–260.

1991b General Laws of Nature and the Uniqueness of the Universe. In: E . Agazzi and A. Cordero (Eds.), *Philosophy and the Origin and Evolution of the Universe*. Dordrecht: Kluwer, 341–360. Repr. in: Scheibe 2001, 276–288.

1991c Erwin Schrödinger und die Philosophie der Physiker. *Zeitschrift für Wissenschaftsforschung* 6, 9–19. English version: Erwin Schrödinger and the Philosophy of the Physicists. In: J. Götschl (Ed.), *Erwin Schrödinger's World View. The Dynamics of Knowledge and Reality*. Dordrecht: Kluwer 1992, 25–34. Repr. in: Scheibe 2001, 108–118.

1991d J. v. Neumanns und J. P. Bells Theorem. Ein Vergleich. *Philosophia naturalis* 28 35–53. Engl. transl.: J. von Neumann's and J. P. Bell's Theorem. A Comparison. In: Scheibe 2001, 419–433.

1991e EPR-Situation and Bell's Inequality. In: W. Spohn et al. (Eds.), *Existence and Explanation. Essays Presented in Honor of K. Lambert*. Dordrecht: Kluwer, 115–129. Repr. in: Scheibe 2001, 434-444.

1991f Substances, Physical Systems, and Quantum Mechanics. In: G. Schurz und G. Dorn (Eds.), *Advances in Scientific Philosophy. Essays in Honor of P. Weingartner on the Occasion of the 60th Anniversary of his Birthday*. Amsterdam: Rodopi, 215–230. Repr. in: Scheibe 2001, 261–275.

1992a The Role of Mathematics in Physical Science. In: J. Echeverria et al. (Eds.), *The Space of Mathematics. Philosophical, Epistemological and Historical Explorations*. Berlin: de Gruyter, 141–155.

1992b Albert Einstein: Theorie, Erfahrung, Wirklichkeit. *Heidelberger Jahrbücher* XXXVI, 121–138 (farewell lecture Heidelberg]. Engl. transl.: Albert Einstein: Theory, Experience, Reality. In: Scheibe 2001, 119–135.

1992c The Divorce Between the Sciences and the Humanities. *Journal of Indian Council of Philosophical Research* IX, 137–150. Repr. in: G. Meggle und U. Wessels (Eds.), *Analyomen 1. Proc. 1st Conference "Perspectives in Analytical Philosophy"*. Berlin: de Gruyter 1994, 362–378.

1993a Heisenbergs Begriff der abgeschlossenen Theorie. In: B. Geyer et al. (Eds.), *Werner Heisenberg. Physiker und Philosoph*. Heidelberg: Spektrum, 251–257. Engl. transl.: Heisenberg's Conception of a Closed Theory. In: Scheibe 2001, 136–141.

1993b A New Theory of Reduction in Physics. In: J. Earman et al. (Eds.), *Philosophical Problems of the Internal and External Worlds: Essays on the Philosophy of A. Grünbaum*. Pittsburgh: University of Pittsburgh Press, 249–271. Repr. in: Scheibe 2001, 352–368.

1993c C. F. v. Weizsäcker und die Einheit der Physik. *Philosophia naturalis* 30, 126–145. Engl. transl.: C. F. v. Weizsäcker and the Unity of Physics. In: Scheibe 2001, 54–68.

1993d Three Remarks Concerning Bell's Inequality. In: A. van der Merwe et al. (Eds.), *Bell's Theorem and the Foundations of Modern Physics*. Singapore: World Scientific 428–435. Repr. in: Scheibe 2001, 445–452.

1994a Zwischen Rationalismus und Empirismus: Der Weg der Physik. In: F. Fulda and R. -P. Horstmann (Eds.), *Vernunftbegriffe der Moderne. Stuttgarter Hegel-Kongress 1993*. Stuttgart: Klett-Cotta, 73-95. Engl. transl.: Between Rationalism and Empiricism: The Path of Physics In: Scheibe 2001, 69–86.

1994b A Most General Principle of Invariance. In: E. Rudolph und I. -O. Stamatescu (Eds.), *Philosophy, Mathematics and Modern Physics. A Dialogue*. Berlin: Springer, 213–225. Repr. in: Scheibe 2001, 501–512.

1994c On the Mathematical Overdetermination of Physics. In: E. Rudolph und I. -O. Stamatescu (Eds.), *Philosophy, Mathematics and Modern Physics*. Berlin: Springer 1994, 186–199. Revised version: The Mathematical Overdetermination of Physics. In: E. Agazzi and G. Darvas (Eds.), *Philosophy of Mathematics Today*. Dordrecht: Kluwer 1997, 269–285.

1994d Canonical Invariance in Physics. In: U. Majer und H. -J. Schmidt (Eds.), *Semantical Aspects of Space-Time Theories*. Mannheim: Bibliographisches Institut, 195–201.

1995a The Rationality of Reductionism. In: R. Zwilling (Ed.), *Natural Sciences and Human Thought*. Berlin: Springer, 101–109. Repr. in: Scheibe 2001, 369–377.

1995b L'origine du réalism scientifique: Boltzmann, Planck, Einstein. In: M. Panza and J. -C. Pont (Eds.), *Les Ravants et l'épistémologie vers la fin du XIXe siècle*, Paris: Albert Blanchard, 157–172. Original English version: The Origin of Scientic Realism: Boltzmann, Planck, Einstein. In: E. Agazzi and M. Pauri (Eds.), *The Reality of the Unobservable*. Dordrecht: Kluwer 1999, 31–44. Repr. in: Scheibe 2001, 142–155.

1995c Laws of Nature: Generality vs. Coherence. In: F. Weinert (Ed.), *Laws of Nature. Essays on the Philosophical, Scientific and Historical Dimension*. Berlin: de Gruyter, 208–226.

1997a The Mathematical Overdetermination of Physics. In: E. Agazzi and G. Darvas (Eds.), *Philosophy of Mathematics Today*. Dordrecht: Kluwer, 269–285 (revised version of 1994c). Reprint in: Scheibe 2001, 571–583.

1997b Missverstandene Naturwissenschaft. In: R. Enskat (Ed.), *Wissenschaft und Aufklärung*. Opladen: Leske und Budrich, 9-29. Repr. in: B.-O. Küppers (Ed.), *Die Einheit der Wirklichkeit*. München: Fink 2000, 69–87.

1997c Zeit und Physik in der Philosophie Carl Friedrich v. Weizsäckers. In: W. Krohn and K. M. Meyer-Abich (Eds.), *Einheit der Natur—Entwurf der Geschichte*. Begegnungen mit Carl Friedrich von Weizsäcker. München: Hanser, 157–171.

1997d The Problem of Reduction in the Theory of Special Relativity. In: D. Ginev and R. P. Cohen (Eds.), *Issues and Images in the Philosophy of Science. Scientific and Philosophical Essays in Honour of Azarya Polikarov*. Dordrecht: Kluwer, 321–342.

1998a On Limitations of Physical Knowledge. *Philosophia naturalis* 35, 41–57. Repr. in: Scheibe 2001, 289–301.

1998b Wie argumentieren Philosophen? In: R. Enskat (Ed.), *Amicus Plato, magis amica veritas. Festschrift für Wolfgang Wieland zum 65. Geburtstag*. Berlin: de Gruyter 1998. 219–227

1999 Lorenz Krüger über Reduktion und Entwicklung der Wissenschaft. In: W. Carl und L. Daston (Eds.), *Wahrheit und Geschichte*. Göttingen: Vandenhoeck & Ruprecht, 115–127.

Reviews of Erhard Scheibe's Books

(1973) MacKinnon, E. 1975. *Philosophy of Science* 42, No. 1, 96–100.
Petersen, A. 1975. *Isis* 66, No. 2, 293–294.
(1997) Bartelborth, T. 1999. *Erkenntnis* 50, 123–127.
Diederich, W. 1999. *J. Gen. Phil. Sci.* 30, 389–396.
Falkenburg, B. 1997. DIALEKTIK 1997/3, 135–139.
Hüttemann, A. 2000. *International Studies in the Philosophy of Science* 14, 333–336.
Stöckler, M. 1997. *Physikalische Blätter* 53, Nr. 6, 552.
(1999) A. Hüttemann. 2000. *International Studies in the Philosophy of Science* 14, 333–336.
(2001) Hellwig, K. E. 2006. *Zentralblatt Math.* 1088.
Da Costa, N. C. A, and D. Krause. 2005. *Studies in History and Philosophy of Modern Physics* 36, 210–217.
van Dyck, M. 2000 *Philosophica* (Belgium) 65, 101–106.
(2006) Fox, T. 2007. *J. Gen. Phil. Sci.* 38, 201–204.
Görnitz, T. 2006. *Physik Journal* 8, Nr. 12, 58
Urban, M. 2006. *Süddeutsche Zeitung* (7.11.).
Mayer, H. 2007. *Frankfurter Allgemeine Zeitung* (24.9.).

Related works

Drüll, D. 2009. Entry on Erhard Scheibe, in: *Heidelberger Gelehrtenlexikon 1933–1986*, Berlin: Springer, 526–527.

Krüger, L., and B. Falkenburg (eds.). 1995. *Physik, Philosophie und die Einheit der Wissenschaften.* Festschrift für Erhard Scheibe. Heidelberg: Spektrum Akademischer Verlag. With contributions of M. Arsenijevic, N. Cartwright, C. Chevalley, L. Daston, W. Diederich, J. Ehlers, B. Falkenburg, A. Grünbaum, M. Hallett, B. -O. Küppers, L. Krüger, K. Lambert, G. Ludwig, P. Mittelstaedt, C. -U. Moulines, F. Mühlhölzer, G. L. Pandit, M. Stöckler, G. Süßmann, and a bibliography of Erhard Scheibe's work until 1994.

Author Index

E. Scheibe, *The Reduction of Physical Theories*, Fundamental Theories of Physics 207,
https://doi.org/10.1007/978-3-662-65000-4

Subject Index

E. Scheibe, *The Reduction of Physical Theories*, Fundamental Theories of Physics 207,
https://doi.org/10.1007/978-3-662-65000-4

Printed by Printforce, the Netherlands